PLANETARY JUSTICE

Stories and Studies of Action,
Resistance and Solidarity

Edited by
Michele Lobo,
Eve Mayes and Laura Bedford

First published in Great Britain in 2024 by

Bristol University Press
University of Bristol
1-9 Old Park Hill
Bristol
BS2 8BB
UK
t: +44 (0)117 374 6645
e: bup-info@bristol.ac.uk

Details of international sales and distribution partners are available at bristoluniversitypress.co.uk

Editorial selection and matter © Michele Lobo, Eve Mayes, Laura Bedford; © Chapter 5 Lowell Hunter and Michele Lobo; © Interstice 2 The Gesturing Towards Decolonial Futures Collective; individual chapters © their respective authors, 2024

The digital PDF and EPUB versions of this title are available open access and distributed under the terms of the Creative Commons Attribution-NonCommercial-ShareAlike 4.0 International licence (https://creativecommons.org/licenses/by-nc-sa/4.0/) which permits adaptation, alteration, reproduction and distribution for non-commercial use without further permission provided the original work is attributed. The derivative works must be licensed on the same terms.

British Library Cataloguing in Publication Data
A catalogue record for this book is available from the British Library

ISBN 978-1-5292-3529-6 paperback
ISBN 978-1-5292-3530-2 ePub
ISBN 978-1-5292-3531-9 ePdf

The right of Michele Lobo, Eve Mayes and Laura Bedford to be identified as editors of this work has been asserted by them in accordance with the Copyright, Designs and Patents Act 1988.

All rights reserved: no part of this publication may be reproduced, stored in a retrieval system, or transmitted in any form or by any means, electronic, mechanical, photocopying, recording, or otherwise without the prior permission of Bristol University Press.

Every reasonable effort has been made to obtain permission to reproduce copyrighted material. If, however, anyone knows of an oversight, please contact the publisher.

The statements and opinions contained within this publication are solely those of the editors and contributors and not of the University of Bristol or Bristol University Press. The University of Bristol and Bristol University Press disclaim responsibility for any injury to persons or property resulting from any material published in this publication.

Bristol University Press works to counter discrimination on grounds of gender, race, disability, age and sexuality.

Cover design: Andrew Corbett
Front cover image: © Lowell Hunter https://saltyone.com.au/

Contents

List of Figures and Tables		v
List of Abbreviations		vi
Notes on Contributors		viii
Acknowledgements		xv
1	Earth Unbound: Situating Climate Change, Solidarity and Planetary Justice *Michele Lobo, Eve Mayes and Laura Bedford*	1
PART I	**Solidarity as Responsibility, Resurgence and Regeneration**	
2	Waking up the Snake: Ancient Wisdom for Regeneration *Anne Poelina, Bill Webb, Sandra Wooltorton and Naomi Joy Godden*	25
3	Farmers as Allies Towards Ecological Justice: Lessons from Water Markets, Colonialism and Theft in Australia's Murray-Darling Basin *Alexander Baird*	39
4	Freshwater Access, Equity and Empowerment in the Indian Sundarban Region *Anwesha Haldar, Kalyan Rudra and Lakshminarayan Satpati*	54
5	Climate Change and Oceanic Responsibilities: Listening and Dancing with Saltwater Country, Australia *Lowell Hunter and Michele Lobo*	71
Interstice 1:	Saturated Strands of (In/Re)Surgent Solidarity *Yin Paradies*	87

PART II	**Solidarity without Borders**	
6	Asserting Indigenous Self-Determination and Climate Justice Through Resisting Coal: A Global North-South Comparison *Ruchira Talukdar*	97
7	Popular Intellectuals, Social Movement Frames and the Evolution of the Anti-Mining Movement in the Niyamgiri Mountains, Odisha, India *Souvik Lal Chakraborty and Julian S. Yates*	122
8	Solidarity as Praxis in Class Struggle *Laura Bedford*	140
Interstice 2:	The Gifts of Failure *The Gesturing Towards Decolonial Futures Collective*	152
Interstice 3:	Face to Face with the Supercyclone Amphan – Kolkata, 20 May 2020 *Sanjana Dutt*	155
PART III	**Learning and Living with Climate Change as Situated Solidarity**	
9	Planetary Justice and Decolonizing Pedagogy: Teaching and Learning in Solidarity with Country *Aleryk Fricker*	163
10	Towards Transformative Social Resilience: Charting a Path with Climate-Vulnerable Communities in the Indian Sundarbans *Jenia Mukherjee, Amrita Sen, Kuntala Lahiri-Dutt and Aditya Ghosh*	178
11	Profane Knowledge, Climate Anxiety and the Politics of Education *Callum McGregor, Beth Christie and Marlies Kustatscher*	196
12	White Audacity and Student Climate Justice Activism *Natasha Abhayawickrama, Eve Mayes and Dani Villafaña*	213
Interstice 4:	Soil Geopolitics and Research as Ecological Praxis *Robin Bellingham*	232
Postscript:	The Earth Is Undone *Alicia Flynn*	242
Index		244

List of Figures and Tables

Figures

4.1	Location map of the Indian Sundarban Region	56
4.2	Community Development Block map of the Indian Sundarban Region	57
5.1	Connection to Saltwater Country	75
5.2	Whale Dreaming	77
5.3	Citizens' Protection Declaration	78
5.4	Reinvigorate	79
5.5	Saltwater Healing	80
5.6	Video artwork	82
10.1	Map of Kumirmari	186

Tables

4.1	Media representations and ground realities of water access in the Sundarban region	61
10.1	Multiple stages of K2A project execution	188
10.2	Summary of the SWOT analysis	189

List of Abbreviations

ADDIE	Analysis-Design-Development-Implementation-Evaluation Model
BIPOC	Black, Indigenous and People of Colour
CDBs	Community Development Blocks
CEDAW	Convention on the Elimination of All Forms of Discrimination against Women
COP	United Nations Conference of the Parties
COVID-19	Coronavirus Disease 2019
CRPF	Central Reserve Police Force
DEECA	Victorian Department of Energy, Environment and Climate Action
ECJ	Education for Climate Justice
FPIC	Free, Prior and Informed Consent
FRA	Scheduled Tribes and Other Traditional Forest Dwellers Act 2006
GBM	Ganga Brahmaputra-Meghna-Delta Region
GORCAPA	Great Ocean Road Coast and Parks Authority
GoWB	Government of West Bengal
GTDF	Gesturing Towards Decolonial Futures
ICAR	Central Inland Fisheries Research Institute
ILUA	Indigenous Land Use Agreement
IPLCs	Indigenous Peoples and Local Communities
ISR	Indian Sundarban Region
KISS	Kalinga Institute of Social Science
KMA	Kolkata Metropolitan Area
LfS	Learning for Sustainability
MDB	Murray-Darling Basin
MGNREGA	Mahatma Gandhi National Rural Employment Guarantee Act
MoEF	India's Ministry of Environment and Forests
MoU	Memorandum of Understanding
MTPA	Million Tonnes Per Annum
NTA	Australia's Native Title Act 1993

LIST OF ABBREVIATIONS

OWW	One-World World
PHED	Public Health and Engineering Department
PML	Proposed Mining Lease
PVTG	Particularly Vulnerable Tribe Groups
RDA	Australia's Racial Discrimination Act 1975
RO	Reverse Osmosis purification techniques
RSMC	Regional Specialized Meteorological Centre
RWH	Rainwater Harvesting
SAPs	Situated Adaptive Practices
SBR	Indian Sundarban Biosphere Reserve
SDB	Sundarban Development Board
SDG	Sustainable Development Goal
SJSM	Sundarban Jana Sramajibi Mancha
SMART	Specific-Measurable-Attainable-Relevant-Timebound
SOPEC	Southern Ocean Protection Embassy Collective
SS4C	School Strike 4 Climate
STR	Sundarban Tiger Reserve
SWOT	Strengths, Weaknesses, Opportunities and Threats
TTF	Teach the Future
UNDRIP	United Nations Declaration on the Rights of Indigenous People
UNFCCC	United Nations Framework Convention for Climate Change
W&J	Wangan and Jagalingou

Notes on Contributors

Natasha Abhayawickrama is an undergraduate university student at the University of New South Wales with over five years of experience as a community organizer. She is currently employed as a part-time grassroots mentor and organizer with the Australian Youth Climate Coalition and as a Research Associate at Deakin University, Australia. Previously as a high school student, she was an organizer with School Strike for Climate (Sydney and national organizer). She is a member of Sapna South Asian Climate Justice Solidarity.

Alexander Baird is Senior Research Associate at Deakin University. His areas of research interests and expertise include green criminology, environmental crime and water theft. He is particularly interested in state and corporate commodification and exploitation of nature, and explores the complex and intricate dimensions of environmental harm from a criminological approach. He is currently working on an Australian Research Council Discovery Project entitled 'Preventing Water Theft in the Murray Darling Basin'.

Laura Bedford is Senior Lecturer in the School of Social and Political Sciences at the University of Melbourne. She is recognized for her critical scholarship across several subfields of criminology, including the interdisciplinary fields of political ecology, crimes of the powerful and state-corporate crime, and policing and crime prevention. She applies her skills and experience in mixed-methods and participatory research, policy analysis and programme evaluation to investigate the complex social and ecological justice challenges and opportunities of the 21st century. She collaborates across disciplines to enhance the interaction between critical basic and applied research in the broad and expanding field of green criminology.

Robin Bellingham is Senior Lecturer in Education, Pedagogy and Curriculum at Deakin University in Melbourne. Her research and teaching focus on how education, pedagogy and research methodologies can respond ethically to the most pressing problems of our time, including educational and political disempowerment and disengagement, the ongoing effects of colonization, and ecological crisis. She draws on posthumanism, decolonial theory, critical theory, critical place inquiry and creative writing genres.

Souvik Lal Chakraborty was Lecturer in Human Geography at Monash University, Australia, where he was also the interim Director of the Master of International Development Practice programme. His research and teaching focus on studies of nature–society relationships. More specifically, his research interests include studies of politics of mining and resource extraction, intersections between Indigenous geographies and political ecology, contentious politics and critical development studies. His doctoral thesis provides an in-depth analysis of an indigenous social movement in India with a specific focus on the eastern state of Odisha.

Beth Christie is Senior Lecturer in Learning for Sustainability and Programme Director for the MSc Learning for Sustainability at the Moray House School of Education and Sport, University of Edinburgh. She is passionate about reimagining and transforming education for a sustainable future. Her research spans policy, academia and professional practice to understand the structural, theoretical and practical processes of, and possibilities for, systemic change.

Sanjana Dutt is a doctoral student at Nicolaus Copernicus University in Poland and pursued her early studies in geography and geomatics in Kolkata, India. With a deep interest in travel and photography, she now concentrates on studying the spatiotemporal dynamics of natural environments. Her ongoing research employs quantitative methods to investigate the complex fragmentation patterns in Poland's forests, particularly due to windthrow events.

Alicia Flynn is a white settler educator, researcher, mother, activist, gardener and novice poet living on the unceded land of the Wurundjeri Woiwurrung people. Based on an inquiry with a public high school in inner-city Narrm Melbourne, her PhD explores the sympoietic relationalities of more-than-human, anti-colonial, climate and place-responsive learning in the precarious times of the Capitalocene. She teaches Environmental Education at a primary school that takes composting seriously and is dedicated to land-responsive ethics and justice in every expression of living, learning and teaching-with the world.

Aleryk Fricker is a proud and sovereign Dja Dja Wurrung man and Lecturer in Indigenous education in the National Indigenous Knowledges Education Research Innovation (NIKERI) Institute at Deakin University, and has worked in the education sector for over a decade. He is an active teacher and researcher and works hard with the next generation of teachers to support them to begin to decolonize their classrooms to allow all students, regardless of their cultural contexts, to access and benefit from First Nations knowledges and pedagogies that support teaching and learning.

Gesturing Towards Decolonial Futures Arts/Research Collective (GTDF) is an interdisciplinary, international, and intergenerational collective of artists, activists, researchers, educators, students and Indigenous knowledge keepers who work closely together to develop public pedagogies and artistic interventions at the interface of two sets of decolonial questions: (1) questions relating to confronting historical, systemic and ongoing social and ecological violence; and (2) questions relating to the unsustainability of modern-colonial systems and ways of being.

Aditya Ghosh is Teaching Fellow with the University of Leicester, UK. He is a human geographer who examines sustainability challenges, socioecological risks and spatial planning amid rapidly changing climatic changes in postcolonial contexts, focusing on different forms of (in)justice and resistance. He deconstructs knowledge hegemonies in global sustainability governance and decolonizes the understanding of sustainable development. He has worked with Universities of Leicester (UK), Heidelberg and Leuphana (Germany), Sussex (UK), Colorado-Boulder (US) and OP Jindal Global (India) as well as with WWF-India and the Centre for Science and Environment as a member of faculty and/or researcher for the past 14 years. In his previous career as a journalist and editor for over a decade, he worked with Encyclopaedia Britannica, *The Guardian*, *The Times of India* and the *Hindustan Times*. He is a recipient of prestigious global awards such as the Chevening fellowship (UK) and the DAAD fellowship (Germany).

Naomi Joy Godden is a *wadjela* (non-Indigenous) woman who was raised and continues to live in Wardandi Country in southwest Western Australia. She is an Australian Research Council DECRA Fellow at the Strategic Research Centre for People, Place and Planet at Edith Cowan University in Bunbury, Western Australia. She is an activist researcher and collaborates with grassroots communities and organizations to undertake Feminist Participatory Action Research for climate justice in the southwest, across Australia and in the Asia Pacific.

Anwesha Haldar is Assistant Professor of Geography at East Calcutta Girls' College, Kolkata. After completing her undergraduate degree in geography at Loreto College, she did a PhD on Sustainable Development in the Indian Sundarban Region at the University of Calcutta. She was also a University Grants Commission Senior Research Fellow at the Department of

Geography, University of Calcutta. She has published many articles in reputed journals, edited books and presented papers on urban geomorphology, climatology and environmental geography in India and overseas.

Lowell Hunter, a proud Nyul Nyul saltwater man from the Kimberley region in Western Australia, grew up on Gunditjmara Country in Warrnambool and now lives on Wadawurrung Country, Geelong, Victoria. Each of these special places has kept him strongly connected to the ocean his whole life. He creates sand art and uses drone photography to capture his work within breathtaking beachscapes. Using only his feet, he carves stories in the sand using the same foot movements he was taught through traditional dance movements his people have practised for countless generations. His artworks tell stories of family, identity and earthly connections.

Marlies Kustatscher is Senior Lecturer in Childhood Studies at the Moray House School of Education and Sport, University of Edinburgh, and a member of the Childhood and Youth Studies Research Group and the Race and Inclusivity in Global Education Network (RIGEN). Her research interests include childhood and intersectionality, children's rights and participation, and participatory and arts-based research methodologies with children and young people.

Kuntala Lahiri-Dutt is Professor at the Crawford School of Public Policy, ANU College of Asia and the Pacific, Australian National University. She has extensively researched the social and ecological politics of natural resources since 1993–1994, initially in India and later on in other Asian countries such as Indonesia, Lao PDR and Mongolia. She has published on the livelihoods of poor and immigrants living on ecological boundaries of land and water, on chars or river islands in her book *Dancing with the River* (Yale University Press, 2013, co-authored with Gopa Samanta).

Michele Lobo is an Australian cultural geographer. Her research agenda invigorates debates on racial and planetary justice informed by diverse Indigenous, Southern, Eastern, Black, Diasporic and Islamic cosmologies of co-belonging. Her knowledge about justice and belonging has emerged through her encounters with Indigenous peoples, ethnic/ethnoreligious minority migrants, refugees, asylum seekers and international students in cities that include Melbourne, Darwin, Sydney, Detroit, Paris, Kolkata and Mumbai. She is editor of *Social & Cultural Geography*, reviews editor of *Postcolonial Studies* and Council Member of the Institute of Australian Geographers. She has published more than 60 scholarly pieces, including three books supported by three prestigious national grants.

Eve Mayes is Senior Research Fellow and Senior Lecturer at Deakin University in the School of Education (Research for Educational Impact). She is currently undertaking the Australian Research Council (ARC) Discovery Early Career Fellowship (DECRA) project 'Striking Voices: Australian School-Aged Climate Justice Activism' (2022–2025). Her PhD (2016) was a participatory ethnographic study of student voice in school reform processes. Her book *The Politics of Voice in Education* was published in 2023 by Edinburgh University Press.

Callum McGregor is Lecturer in Education at the Moray House School of Education and Sport, University of Edinburgh, where he teaches the MSc Social Justice and Community Action and the MA Learning in Communities courses. His research interests, located at the intersection of social movement activism and education, continue to shape his current work on education for climate justice.

Jenia Mukherjee is Associate Professor in the Department of Humanities and Social Sciences at the Indian Institute of Technology Kharagpur, India. She is interested in transdisciplinary approaches, more specifically political ecology and environmental humanities. It is her conviction that 'knowing' and 'doing' matter simultaneously. She is committed to nurturing academic social responsibility by closely collaborating with practitioners, government organizations, think tanks and user groups. She has been the recipient of Australian Leadership Awards (2010, 2015), the World Social Science Fellowship (2013), the Rachel Carson Writing Fellowship (2018–2019) and the Nippon Foundation Fellowship (2020, 2022). She is currently leading several global partnership projects, generously funded by the Social Sciences and Humanities Research Council (Canada), DAAD (Germany), the Swiss National Science Federation (Switzerland) and the International Water Management Institute (Srilanka) on climate resilience and coastal fisheries in the Global South.

Yin Paradies is a Wakaya man who is Professor and Chair in Race Relations at Deakin University, Australia. He conducts research on the health, social and economic effects of racism, anti-racism theory, policy and practice, as well as on Indigenous knowledges and decolonization. He is an animist anarchist activist committed to interrupting the devastating impacts of modern societies. He seeks mutuality of becoming and embodied kinship with all life through transformed ways of knowing, being and doing grounded in wisdom, humility, respect, generosity, downshifted collective sufficiency, voluntary simplicity, frugality, direct participation and radical localization.

Anne Poelina is a Nyikina Warrwa woman from the Kimberley region of Western Australia. She is Professor and Chair Indigenous Knowledges and Senior Research Fellow at the Nulungu Institute Research University of Notre Dame, and is Adjunct Professor at the College of Indigenous Education Futures, Arts & Society, Charles Darwin University, Darwin. She is the Murray Darling Basin (MDB) inaugural First Nations appointment to its independent Advisory Committee on Social, Economic and Environmental Sciences (2022). She is a member of the Institute for Water Futures and a Visiting Fellow at the Water Justice Hub at the Australian National University, Canberra. She is also the inaugural Chair of the Martuwarra Fitzroy River Council.

Kalyan Rudra is presently Chairman of West Bengal Pollution Control Board and a former member of the Central Pollution Control Board, India. He is a geographer by academic training with specialization in river and water management. Since 2005, he has been an expert member in a committee constituted by the apex court of the State of West Bengal for cleaning the Ganges. He has written five books and many research papers published in reputed journals in India and abroad. He was also an advisor to the International Union for Conservation of Nature.

Lakshminarayan Satpati is Professor in Geography and Director of the University Grants Commission-Human Resource Development Centre at the University of Calcutta, India. He has supervised 26 PhD researchers and two postdoctoral fellows. He is a member of the West Bengal State Action Plan for Climate Change, 2021–2030 and is Chairperson of the committee to restructure the National Atlas and Thematic Organisation (NATMO), India. He holds several academic and administrative responsibilities at universities across India and in organizations like the Geographical Society of India, Kolkata, and the Indian Meteorological Society (IMS), Kolkata Chapter.

Amrita Sen is Assistant Professor of Sociology at the Indian Institute of Technology, Kharagpur, India. Her research focuses on the politics of forest conservation, urban environmental conflicts, climate adaptation action plans and ecological resilience. She is the author of *A Political Ecology of Forest Conservation in India: Communities, Wildlife and the State* (Routledge, 2022) and is co-editor of *Regional Political Ecologies and Environmental Conflicts in India* (Routledge, 2022). As a scholar with the Climate Social Science Network, Brown University, she currently investigates funded projects on adaptation and social resilience in climate vulnerable islands of the Indian Sundarbans.

Ruchira Talukdar is Co-Founder and Project Director at Sapna South Asian Climate Solidarity, an Australian-based climate justice network for effective Global North solidarity for just climate futures in the Global South. Her research focuses on comparative aspects of climate justice and climate activism between the Global North and South. Her PhD thesis compared coal conflicts and protest movements in India and Australia, with an emphasis on the intersections between grassroots and Indigenous movements and mainstream environmental activism. She has worked in the environment movement in India and Australia in Greenpeace, the Australian Conservation Foundation and Friends of the Earth for two decades.

Dani Villafaña is an undergraduate student and community organizer. She was a lead organizer with School Strike for Climate and was previously a campaigner with Sweltering Cities. She has worked in the gender justice sexual assault victim-survivor advocacy space as a survivor-advocate, around the legislation of sex education, protections for survivors of gendered violence and access to reproductive healthcare. She currently works as a campaigner with Fair Agenda and as a Research Associate at Deakin University, Australia.

Bill Webb is a Traditional Owner from the Wardandi Noongar language group located in the southwest region of Western Australia. He leads the Wardan Aboriginal Cultural Centre in Injidup.

Sandra Wooltorton is a multigenerational Australian of Anglo-Celtic heritage, born and raised in Noongar Country, southwest Western Australia and living much of her life in the Kimberley. She is a Professor and Senior Research Fellow at the Nulungu Research Institute, at the University of Notre Dame Australia, and has published substantial interdisciplinary and transdisciplinary research across fields such as sustainability education, research for social change, Aboriginal education, environmental philosophy and cultural geography. She leads research that applies transitions discourses to Australian contexts. She collaborates on an Australian Research Council (ARC)-funded research project entitled 'Intergenerational Cultural Transfer of Indigenous Knowledges'.

Julian S. Yates is a human geographer with a research career focused on exploring and supporting just approaches to environmental stewardship and development. He has worked with recycler cooperatives and movements in Brazil, Quechua communities in Peru's Southern Andes and First Nations in British Columbia, Canada. He has published on topics such as alternatives to neoliberal development, Indigenous knowledge approaches for Indigenous stewardship and community-led adaptation to climate change.

Acknowledgements

> You are now part of a new story,
> Let's make a stand, let's plan, let's send the Dream out!
> Anne Poelina, Nyikina Warrwa woman,
> the Kimberley region, Western Australia[1]

We acknowledge First Nations people who are storytellers and guardians of sentient, sacred land and sea country in so-called Australia. Our gratitude and respect to Ancestors, Elders and Traditional Custodians of the unceded lands of Wurundjeri Country and Wadawurrung Country in southeastern Australia where we live and work. As editors and members of the Earth Unbound Collective, this Country kindled our dreams for planetary justice. Thanks to Nyul Nyul saltwater man and sand artist Lowell Hunter for the cover image that follows the way of the creator Bunjil, the ancestral wedge-tailed eagle, in reinvigorating land and sea country in Australia and beyond.

This book, and the Earth Unbound Collective, came together amid the co-occurring crises of Australia's Black Summer and the COVID-19 pandemic in early 2020. We thank the Collective, activists and scholar-activists from Australia, Canada, Japan, India, the UK and beyond, who participated regularly in lively debates that have unfolded over the past four years. This book would not have been possible without your provocative stories, questions and participation in the two-day blended symposium held at Deakin University in November 2021. We are glad that many of you were able to contribute to this book and thank you for your refreshing ideas and tireless work! Thanks to the presenters from the Gesturing Towards Decolonial Futures (GTDF) Collective, Scientist Rebellion, Extinction Rebellion, Sapna South Asian Climate Solidarity and other activist/scholar-activist groups.

Special thanks to Yin Paradies and Anna Halafoff who supported Michele in leading the Earth Unbound Collective in 2020 at Deakin University.

---

[1] Chinna, N. and Poelina, A. (in press) *Tossed Up by the Beak of a Cormorant. Poems of Martuwarra Fitzroy River*, Fremantle Press: Fremantle, WA.

Geographers Charlotte Jones and Vicky Zhang: your convivial support, hard work, enthusiasm and laughter were crucial in organizing meetings and the hybrid 2021 international symposium, and setting up the website.

We acknowledge the financial support provided by the School of Humanities and Social Sciences and the Alfred Deakin Institute for Citizenship and Globalisation (ADI), Deakin University, for activities of the Collective over the past four years, including an international symposium in 2021. Thanks to Fethi Mansouri, Emma Kowal, Louise Johnson and Jack Reynolds. Special thanks to research leaders Anita Harris and Maurizio Meloni of the Mobilities, Diversity and Multiculturalism Thematic Stream and Culture, Environment and Science Streams respectively at ADI. We also acknowledge and thank the Deakin Centre for Regenerating Futures for a small grant for the book's copyediting. Thanks to Paula Muraca for your copyediting wizardry and Zoe Brittain for maintaining the web presence of the Earth Unbound Collective in 2022/2023: https://earthunboundcollective.wordpress.com. Eve Mayes acknowledges the support of Research for Educational Futures (REDI) at Deakin University and the Australian Research Council (ARC) (Discovery Early Career Research Award; project number DE220100103). The editors thank the Deakin University Office of the Deputy Vice-Chancellor Research and Research for Educational Futures (REDI) for enabling this book to be Open Access.

Thank you, Emily Watt, Anna Richardson and the team at Bristol University Press. We hope this book materializes collective political and spiritual dreams of thinking, listening, inhabiting and acting with the planet.

1

Earth Unbound: Situating Climate Change, Solidarity and Planetary Justice

Michele Lobo, Eve Mayes and Laura Bedford

The seeds of this book were sown by the Earth Unbound Collective, a collective of scholars and activists that formed amid the accelerating crises of anthropogenic global warming, catastrophic 'natural' disasters and struggles for climate justice and fossil-free energy futures. In late January 2020, we came together in the wake of the 'Black Summer' bushfires, which had been ignited by the hottest and driest summer ever recorded in Australia (Lobo et al, 2021). Fires were raging across multiple southeastern states, including our home in Victoria, and a national emergency had been declared (Commonwealth of Australia, 2020). Beyond Australia, fires burned across the Amazon Rainforest and the Pacific coast of the US, floods devastated Uganda, Burundi and Pakistan, and supercyclones pummelled coastal areas of India, Bangladesh, New Zealand and Vanuatu.

As humans continue to agitate and reshape planetary forces, this book responds to the social, ecological and ethical challenges of climate change. We write this opening chapter and edit this collection as three non-Indigenous Australian women working across the diverse disciplines of geography (Lobo), education (Mayes) and criminology (Bedford). We centre 'planetary justice' as a provocation to unsettle human exceptionalism, climate inaction and the legacies of white colonial domination highlighted by feminist geographers, philosophers of science, Indigenous philosophers, decolonial/postcolonial historians and political theorists (Haraway, 2016; Connolly, 2017; Whyte, 2020a, 2020b; Chakrabarty, 2021; Sultana, 2021a, 2021b, 2022). The first meeting of Earth Unbound took place in person under smoky skies, as air quality in Melbourne was declared the worst in the world. By our second meeting, an international COVID-19 public health emergency had been declared and we were locked down at home. We met on Zoom to consider how these 'overlapping and uneven crises' (Sultana, 2021a, p 447) reinforced local and global inequalities. The uninsured, inadequately housed, communities of colour, those in precarious work and at the frontline of 'clean[ing] the world' (Vergès, 2019, p 1), were left to bear the brunt of the changing climate compounded by the pandemic. The brazen vaccine nationalism that ensued (Byanyima, 2022) reinforced the extent to which the powerful use their capital and influence to further their own interests and exacerbate existing injustices (Sultana, 2021a; Maddrell et al, 2023).

In Australia, as isolated, locked-down and bewildered people waited for big pharma to 'get things back to normal' with a fast-tracked vaccine, the (then) Liberal Party Prime Minister Scott Morrison, infamous for his fondling of coal on the parliamentary floor, called for a 'gas-led recovery' from the economic impacts of the pandemic. This call has subsequently been supported by both major political parties in government: Australia is the fifth largest producer and the second largest exporter of coal, and the world's largest exporter of liquefied gas (Geoscience Australia, 2022). While writing this chapter in late 2023, fracking in the massive 2.8 million-hectare Beetaloo Basin onshore shale gas field in the Northern Territory was approved (Parliament of Australia, 2023b), despite lip-service commitments to net zero and fierce opposition from Indigenous groups, environmental activists, pastoralists and scientists (Fitzgerald, 2023). In 2022, young climate advocate Anjali Sharma joined forces with Senator David Pocock for a legislative campaign for a statutory duty of care to protect Australian children from climate change harm (Duty of Care and Intergenerational Climate Equity Bill, Parliament of Australia, 2023a).

At the end of 2022, the in(action) emerging from the cold and bureaucratic spaces of the United Nations Conference of the Parties (COP 27) at Sharm El Sheikh, Egypt, evidenced the power and influence of the fossil fuel lobby over grassroots calls for climate justice (Klein, 2022). The Earth Unbound Collective came together to consider the international failure to address the intersecting crises of the pandemic, rising inequalities, climate change and biodiversity loss. We shared frustration at the hollowing out of an international response, which, through inaction, served to fuel and intensify the multiple intersecting injustices of colonialism, white supremacy, patriarchy, capitalism and ecological debt (Salleh, 2017). We discussed alternative perspectives to those emanating from the endless summits and their greenwashing and technocratic approaches to 'fixing' the crises that strengthen the authority and legitimacy of the status quo. For those, like us, whose lives were hitherto undisturbed (at least seriously or directly) by the devastating consequences of infectious disease and climate change, the 'unprecedented' waves of viral-ecological events in 2020 seemed apocalyptic. But as Potawatomi philosopher Kyle Powys Whyte emphasizes, Indigenous and racialized peoples have long endured shattering violence and destruction from 'different forms of colonialism: ecosystem collapse, species loss, economic crash, drastic relocation, and cultural disintegration' (2018, p 226). Northern forms of environmentalism still struggle to attune to these dynamics of intersectional injustice.

The Collective seeks alternative modes of responsibility and stories of solidarity that challenge dominant assumptions of the 'climate movement'. For example, we discussed the challenge levelled by the Wretched of the Earth Collective – a collective of Indigenous, Black, Brown and diaspora activist groups and scholars – in their open letter to the activist group Extinction Rebellion in 2019. The Wretched of the Earth draw attention to the inextricable entanglements between climate change, colonialism and capitalism, and the limitations of Northern approaches to climate activism. Referencing school strike activist Greta Thunberg's 'Our house is on fire' speech to the World Economic Forum in 2019, they write:

> For many of us, the house has been on fire for a long time: whenever the tide of ecological violence rises, our communities, especially in the Global South are always first hit. We are the first to face poor air quality, hunger, public health crises, drought, floods and displacement ... Our communities have been on fire for a long time and these flames are fanned by our exclusion and silencing. (Wretched of the Earth Collective, 2019, paras 4 and 6)

Since colonization, First Nations peoples have led intergenerational struggles *against* environmental racism and inequality, and *for* sovereignty, land and water rights, and ecological justice (Nolan, 2019). Amid the devastating experiences of loss, asphyxiation and political inertia that have long characterized global capitalism and colonialism, there remain embodied heterogeneous forces of resistance and potent collective and critical processes of world making that model alternative ways of being, knowing, living and relating (Bawaka Country et al, 2016; Lahiri-Dutt, 2017; Theriault et al, 2019; Sultana, 2022). In Australia, Indigenous youth campaigns for Healing Country and climate justice are mobilized by Indigenous-led groups like Seed Mob, Australia's first Indigenous youth climate network, and Original Power, a community-focused Aboriginal organization for self-determined power solutions. Across the diversity of Pacific Island nations and among the Pacific diaspora in Australia, the Pacific Climate Warriors mobilize to take on the fossil fuel industry, as have Indigenous activists in other former colonies and settler colonial nations (Unigwe, 2019).

As the COVID-19 pandemic unfolded, social movements cross-fertilized and drew connections between various crises, prising out links between the spread of COVID-19, racist violence and resistance, workers' struggles, gender-based violence and climate justice. These struggles continued to move with defiance, power and intergenerational energy against the 'unthinking, unfeeling objectivity' (Stengers, 2020, p xiv) of the science of global warming and the necropolitical sleepwalking stance of neoliberal planetary governance. The Earth Unbound Collective has evolved and grown since 2020, and now has over 100 members from multiple sectors, disciplines, and perspectives from around the world. Following the publication of the IPCC Sixth Assessment Synthesis Report (2023), which highlights increasing climate change impacts, climate-related risks and the need for prioritizing equity, climate justice and social justice, this book asks: how can we respond to these conditions with collective action and solidarity?

Earth Unbound: activism and solidarity

The book aims to shift the scope of critical debates on the 'Anthropocene', 'anthropogenic' climate change, the limits of justice and discourses of catastrophic planetary futures. In this opening chapter, we reflect on emergent debates from within the Earth Unbound Collective. We offer 'planetary justice' as a provocative concept and a hook for the collection to emphasize the differentiated experiences of injustice across the planet, and the need for intersectional, ground-up solidarities that cross national and onto-epistemic borders. This collection asks: 'what challenges and opportunities are there for strengthening solidarity in furthering the goals of climate action and planetary justice?'

The politics and ethics of activism

The chapters in this book are written by the Earth Unbound collective of scholars and activists who continue to struggle for planetary justice. Between 2020 and 2023, participants in the Collective took turns to present in various formats at monthly 'dialogues', sharing stories and struggles across different positionalities, disciplinary backgrounds, political contexts and geographical locations. They spoke and wrote from Broome, Perth, Hobart, Sydney, Melbourne (Australia), Vancouver (Canada), Kolkata (India), Kyoto and Tokyo (Japan), Nairobi (Kenya), Dunedin (New Zealand), Edinburgh (Scotland), Cardiff (Wales) and Manchester (England). They identify variously as First Nations/First Peoples, Black, Brown and People of Colour, white settlers, global citizens, and as scholars and/or as activists. Contributors are differentially positioned within and beyond academia. Some are employed in ongoing positions, others are precariously employed on sessional or short-term contracts, while others are positioned beyond the academy, but strategically partner with academics where this work can be 'of use' and 'in service' to grassroots concerns (see Joseph-Salisbury and Connelly, 2021, pp 57, 87). Over the four-year period of Earth Unbound meetings and of compiling this book, some members of the Collective positioned in Australian universities had to re-apply for their tenured university jobs in 'spill and fill' processes; others did not have their short-term contracts renewed, while others remained in ongoing secure employment. In England and Scotland, some academic members engaged in sustained periods of stop work actions, while in Australia, some began industrial action.

Some members shied away from identifying with the term 'activist' and questioned the idealized and exclusionary framing of this term. Others were already involved in recognizable forms of climate justice activism; in grassroots networks or partnering with organizers in these networks. Others still have questioned the relationship between modes of activism and protest mobilized in privileged spaces, and the resistance or defence taken up by those on the frontline of exploitation and climate change. Some have advocated for engaging directly with the state and current institutional forces, while others argued that this strategy risks bolstering state power and undermining prefigurative and decolonial modes of governance (Simpson and Pizarro Choy, 2023). These thorny questions of defining and delimiting 'activism' and change-making strategies are taken up in a variety of ways throughout the chapters in this book.

For those of us positioned in academia, a key thread running through Earth Unbound is a questioning of our place in the 'neoliberal-imperial-institutionally-racist university' (Joseph-Salisbury and Connelly, 2021, p 1): an institution where the production of knowledge is 'principally governed by the West and for the West', serving to 'reproduce and justify colonial hierarchies' (Bhambra et al, 2018, pp 5–6). We have reflected on the complicities and limits of scholar-activism for those of us still working within what Unangax̂ scholar Eve Tuck and K. Wayne Yang call the 'academic industrial complex', whose 'ethical standards … don't always do enough to ensure that social science research is deeply ethical, meaningful or useful for the individual or community being researched' (2014, p 223). We have pondered the role of scholar-activists in struggles for justice, given the hegemony of the Anglosphere in giving voice to this crisis and the whiteness of climate activism in many contexts. We have questioned how the metricized 'output' demands of the academy are frequently incommensurable, or at least sit 'in tension with', the goals of frontline communities and grassroots struggles (Simpson and Pizarro Choy, 2023, p 3). At the same time, dreams for decolonizing the university persist (Lobo and Rodriguez, 2022).

We have collectively shared strategies to find and exploit what Joseph-Salisbury and Connelly call 'pockets of possibility to (partially) mitigate, offset, and utilise the complicities that arise from affiliating with institutional power' (2021, p 3). Conversations within the Collective have interrogated the role of a scholar(-activist), the privilege of the 'ivory tower', what solidarity with grassroots struggles looks like and what 'useful' research might look like. Through reading, sharing, learning, unlearning, relearning and writing, Earth Unbound has engaged, as a collective, with multiple perspectives on climate change and planetary thought, privileging generosity, humility and interdependence.

Solidarity: a polyphonic praxis

The naming of the Collective as Earth Unbound draws attention to the openness of planetary thought, feeling and action that provides possibilities for nourishing solidarity across difference in human and more-than-human worlds. Although Bruno Latour (2018) invites us to be earth bound in the current 'climatic regime', in 'unbinding', we also ardently resist separation of climate justice from the entangled injustices of colonialism, capitalism, patriarchy, biodiversity loss and extinction. The Earth is not 'unbound' *yet*: when climate change displaces people from their homelands, they encounter the enforcement of borders and carceral geographies that bind up and inhibit movement (Walia, 2021). Considering that the unequal impacts and vulnerabilities of climate change operate at a planetary scale, conversations within the Collective have focused on how diverse perspectives of human and more-than-human justice can best be mobilized in action to strengthen solidarity across borders that are material, elemental, political, disciplinary, institutional, interspecies and intersectional. Our collective vision of the planet centres as well as decentres the human, and critically engages with key concepts such as the Anthropocene, Country, activism, solidarity and justice, and their relationship with each other. We situate this book's figuration of speaking and acting with the planet within the broader context of the turn to planetary social thought, the planetary age and planetary imaginaries across the humanities and social sciences. We recognize the discomfort but also the possibilities of the theoretical shift to the planetary amid the everyday violence of climate change (Yusoff, 2018; Chakraborty, 2021).

A key issue for the contributors of this book is the possibility of solidarity, the 'politics of solidarity' (Land, 2015) and enactments of solidarity across intersectional differences, interwoven histories and geographies of genocide and ecocide (Pugliese, 2021), and incommensurable paradigms of ethics and justice that are 'fluid and unsolvable' (Curley et al, 2022, p 1052). Indeed, in the (so-called) 'environmental' movement, there have long been 'contradictions of solidarity', as Curnow and Helferty name them, where environmental activist spaces have been 'default white space[s]' (2018, p 146). How can there be solidarity without acknowledgement of the environmental movement's co-implication in continuing structural injustices? In this book, as in Earth Unbound meetings, we do not seek to draw consensus or generate a sort of groupthink around the complex, situated and differentiated perspectives related to theorizing and responding to climate change (in)justices, but nor do we turn away from these complexities.

Contributors to Earth Unbound, and to this book, have differed in their emphases on what the central 'problem' is, and what an international and collective response should look like. These differences are reflected in the language and concepts used across this book. Some concepts are contentious in usage, or are used differently, or are incorporated with greater or lesser emphasis. For example, some contributors use the language of Global North and the Global South to understand spatially differentiated racial and knowledge hierarchies with a terrestrial bias, notwithstanding acknowledgement that they are 'inherently awkward' terms (Hunter 2020, p 1239; Lahiri-Dutt, 2017). Others problematize the terms 'Global North' and 'Global South' as skirting class in discussing global inequality, arguing that 'North' and 'South' are vague and inaccurate geographic boundaries that break down the possibilities of global class solidarity. Some speak of 'environmental justice' and 'ecological justice', while others use 'climate justice' or 'planetary justice', and yet others reject 'justice'. Some foreground colonialism, postcolonialism and/or decoloniality and the concepts of 'oppressed', 'suppressed' and 'minoritized', while others foreground capitalism, class relations and hegemony.

Within Earth Unbound, we have been increasingly comfortable with our incommensurable epistemologies over time. We are much more interested in noticing who speaks for whom, who is considered to have the power to give voice and say what knowledge is, rather than being 'right'. As editors, we make explicit these differing inflections and articulations of activism, solidarity and justice, and the diverse ontologies that underlie the chapters in this book, including Indigenous, materialist, poststructural and pragmatic. Rather than draw consensus, flatten difference or smooth over sites of apparent contradiction, we foreground that this book intends to bring these multiple vocabularies of climate crisis, planetary collapse, activism, solidarity and justice into productive relation, in frictive encounters (see Puar, 2012). We seek to avoid losing sight of the distinctive power of each of these stories and studies by trying to reconcile these differences. We are inspired by Tuck and Yang's account of an 'ethic of incommensurability' as an 'alternate mode of holding and imagining solidarity' (2018, p 2).

Acknowledging the possibility of diverging pathways across paradigms 'means that we cannot judge each other's justice projects by the same standard, but we can come to understand the gap between our viewpoints, and thus work together in contingent collaboration' (Tuck and Yang, 2018, p 2). Working together 'in contingent collaboration' does not flatten out or gloss over differences, but holds them together generatively. However, our 'dissonant polyphony' is a 'plurilogue' that links 'different yet co-implicated constituencies and arenas of struggle' (Shohat, 2001, p 2). Plurilogues *pursue* rather than gloss over dissimilarities and incommensurabilities, relationally positioning 'different patterns/operations of power ... like the angling of mirrors in a kaleidoscope' (Roshanravan, 2014, p 43). It is the hope of this book that these polyphonic accounts will prompt further challenging conversations about what a praxis of 'planetary justice' that is political, ethical, decolonial and *unbound* might look like across widely different contexts. Rather than adopt a homogenizing framework, the chapters in this book highlight injustice and 'justice' in myriad forms; across human difference, generations, species, life/nonlife and deep time. In the rest of this opening chapter, we further contextualize the concepts of injustice and planetary justice that animate the book, before outlining the three parts into which this book is divided.

The many meanings of (in)justice

From the outset, the intention of Earth Unbound was to be more than an academic 'talk fest', more than another university 'network' or 'incubator' of inaccessible research 'outputs', or an avenue for the university to attract research funding. Simultaneously, we have countered the dismissal of 'talk' of social injustices as irrelevant to climate change activism, such as the statement by Jonathan Morgan of Extinction Rebellion in 2020 in an interview in *Vice*: 'I can't say it hard enough. We don't have time to argue about social justice' (Dembicki, 2020; cited and critiqued by Whyte, 2020a, p 56). Instead, we agree with Kyle Powys Whyte that such 'epistemologies of urgency', rushing to renewable and activist 'solutions', without addressing underlying colonialist and capitalist logics of action, risk repeating 'the moral wrongs and injustices of the past' (2020a, p 56). Past and present injustices must be named and reckoned with.

While the original Earth Unbound Collective came together around climate change, questions of injustice rapidly emerged as central to the Collective's considerations of what it means to be 'Earth bound' and what 'unbinding' entangled structures of oppression might look like, in scholarship, in action or otherwise. We have not sought to reach a consensus about what constitutes the foundational injustices structuring the climate crisis. Indeed, in Interstice 1, Yin Paradies questions whether the 'modern(ist) concept' of justice knows 'how to meet the ghosts (Akomolafe, 2021), wraiths, spectres, shades, and hauntings of history'.

And yet, for many contributors, the injustices intensified by climate change need to continue to be identified and addressed. These are not only the inequitably distributed effects of the climate crisis, but also the injustices of impunity and who is named as responsible for the climate crisis. The 'Anthropocene' has become a ubiquitous term to describe how the geological record 'speaks' of the impacts of human extraction and burning of carbon-rich fossil fuels, deforestation, industrialization and intensive large-scale agriculture over a relatively short period of time (Crutzen and Stoermer, 2000). However, the discursive violence of grouping together a general, universal 'Anthropos' as responsible for these changes in the geological strata has been repeatedly called out. For example, Kathryn Yusoff has called for greater specificity about 'what and who gets marked in Anthropocene origin stories' (2018, p 28) and Macarena Gómez-Barris argues that there can be no universalizing idiom and viewpoint term of the 'Anthropocene' that addresses ' "humanity" as a whole' (2017, p 4).

Across this book, contributors intertwine discussion of the injustices of colonialism and capitalism as drivers of these changes in the geological strata. Following Yellowknives (Weledeh) Dene First Nation scholar Glen Sean Coulthard (2015), colonialism is not a matter of the past, but of the present and past, and present capitalism and colonialism are co-dependent. Dispossession is 'the theft not only of the material of land itself, but also a destruction of the social relationships that existed prior to capitalism' and is an 'ongoing feature of the reproduction of colonial and capitalist social relations in our present' (Coulthard, 2015, n.p.). Jason Moore uses the term 'Capitalocene', not as a 'new word to mock the Anthropocene', but rather as:

> a way of making sense of the planetary inferno, emphasizing that ... [t]he climate crisis is a geohistorical moment that systemically combines greenhouse gas pollution with the climate class divide, class patriarchy, and climate apartheid. The history of justice in the twenty-first century will turn on how well we can identify these antagonisms and mutual interdependencies, and how adeptly we can build political coalitions that transcend these planetary contradictions. (2019, p 54)

How might alternative, pluralist versions of 'justice' be forged across sometimes diverging onto-epistemologies and uneven geographies and histories? The following section outlines the emergence of planetary justice as a concept, our understanding of it, and its possibilities and perplexities.

Towards planetary justice

This book responds to the urgent need for capacious frameworks of justice that can grapple with the historical and contemporary *in*justices that are inextricable from the climate emergency (Schlosberg, 2014; Tschakert et al, 2020). Contributors draw on Indigenous, Black, Southern, ecosocialist and ecofeminist critiques of Western liberal philosophies of environmental and climate justice, and respond to calls for more relational, radical and decolonial solidarities framings of justice (Bulkeley, 2013; Parsons et al, 2021; Sultana 2021a, 2021b; Whyte, 2020a). Contributors call for a mode of justice that is *unbound* from normative frameworks and spatial constraints, that builds up from situated and felt 'injustice, indignation, and harm' (Barnett, 2010, p 252), and that regenerates and heals the Earth.

This book's focus on planetary justice braids diverse perspectives, a pluriversal politics of the possible, and a planetary ethics that might better address injustice in these times (Ruddick, 2017; Derickson, 2018; Clark and Szerszynski, 2020; Escobar, 2020; Chakrabarty, 2021; Schmidt, 2022; Spivak, 2003). We are inspired by Arundhati Roy (2020), who perceived the rupture of normality produced by the COVID-19 pandemic in terms of despair, but also as a portal to 'rethink the doomsday machine we have built'. The climate change machine also offers these opportunities. Following Roy (2020), we can leave behind 'dead ideas', tell the truth about the injustices that structure the past and present, 'imagine another world', and fight for justice for and beyond humanity that is yet to come.

Environmental justice, ecological justice, climate justice and multispecies justice are concepts accompanied by contention that offer significant possibility. The chapters in this book draw on these concepts and refresh them through Global North/South dialogues and grounded participatory action. The concept of planetary justice is indebted to concepts and praxes of environmental justice, climate justice, ecosocialism, ecofeminism and the long lineage work of postcolonial scholars theorizing the 'planetary'. Climate justice engages with key concerns of environmental justice that emerged as a movement against environmental racism in the US in the 1960s and that was taken up in decolonial and anti-capitalist movements around the globe (see, for example, Shiva, 1991). Environmental justice focuses on human rights, equitable access to environmental goods and protection from environmental harms. The concept of climate justice is entangled with longstanding environmental justice, ecosocialist and ecofeminist demands, and calls for greater 'inclusion, autonomy, transparency, compensation and sustainability' (Schlosberg, 2014, p 10). Climate justice therefore critiques the unequal contribution to, and burden of, climate change around the globe between different regions, countries and communities, gender inequalities and environmental racisms, as well as highlights the wisdom of First Nations peoples (Jafry et al, 2018). Today, global environmental *in*justice, ecological and climate debt continue to be central to conceptualizations of global solidarity and to demands for climate reparations from the economic core to the economic periphery (Hornborg and Martinez-Alier, 2016). Although climate justice demands make a more assertive call for the transfer of resources for mitigation, adaptation and reparation from the privileged to the most vulnerable (Táíwò, 2022), tensions emerge when justice principles are framed by academic theories and elite nongovernmental organizations (NGOs) rather than grassroots movements and local communities (Scholsberg, 2014).

Māori-led scholarship by Parsons et al (2021) critiques narrow Western liberal philosophies of environmental justice that focus solely on distributive equity, procedural inclusion and the recognition of cultural difference (see, for example, Fraser, 1997; Fraser and Honneth, 2014). They call for a greater emphasis on Indigenous ontologies and epistemologies that attune to intergenerational and more-than-human justice. Bulkeley et al (2013) agree that discursive framings of climate justice that position nation states within distributive and procedural paradigms fail to consider political possibilities that emerge from local places. The outcome is that distributive paradigms of justice focus on strategies of adaptation (imposition of a duty to devote resources) and mitigation (imposition of a duty to cut back on activities for the common good; a good example of this paradigm is captured in the IPCC Working Group III report, 2023) rather than rights and accepted responsibilities. Procedural justice focuses on 'who should take decisions over what, by what means and on whose behalf', but there is less transparency on the deeply democratic right to participation that includes the most vulnerable (Bulkeley et al, 2013, pp 916–917). The call is therefore for a just climate politics and future research that considers the possibilities and potential challenges of centring the local level. Such framings of *planetary justice* would draw attention to ecological interdependencies that distribute agency more widely and co-compose the plurality of difference that runs through human and more-than-human worlds (Lobo, 2019).

This book considers distributive and postdistributive framings of justice together with diverse experiences of injustice and solidarity that emerge from grassroots struggles for equity, dignity, land rights and living Indigenous/Southern protocols. In discussing environmental justice and climate justice, Whyte (2020a, 2020b) acknowledges the urgency for climate action, but calls for slowing down and greater attention to living Indigenous traditions and kinship relations of consent, trust, accountability and reciprocity that failed to be affirmed by colonialism, capitalism and industrialization. In research informed by empirical work in South Asia, Farhana Sultana (2021a, 2021b) engages with feminist and political ecological scholarship to argue for critical climate justice and 'radical solidarities that can revolutionize thinking on climate change' (2021b, p 1727). This book is profoundly indebted to these streams of thought that engage with and challenge concepts of climate justice, environmental justice and distributive justice.

The climate crisis is undoubtedly of 'planetary scale', necessitating a 'global response' (Simpson and Pizarro Choy, 2023, p 2). Excess carbon dioxide emissions know no borders, and their effects threaten human and more-than-human life across species-level differences and generations, demanding what Sue Ruddick (2017, p 121) calls 'ethical engagement in planetary terms'.

However, those considering how to reduce fossil fuel sources of energy and technosolutionist 'green' responses as a global project must also acknowledge the highly differentiated impact of this shift both within and between nation states (see Satgar, 2022). Planetary justice requires a multiscalar and multitemporal approach that recognizes the imperative to address climate change as an existential threat on a global scale, *and* the localized praxis of responding within unique contexts that climate change entails. As Biermann et al argue, planetary justice 'transforms the older notions of "global" justice with their focus on international institutions, international relations and "global order". Planetary justice, instead, is equally concerned with the global and the local, with state and non-state actors, and with individuals and collectives' (2020, p 1).

Rather than deny the ontological reality of global power structures that will insulate the already privileged – and retreat into the pluralism and the essentialization of the colonial subject as a product of geography and custom instead of history and law (Mamdani, 2012) – planetary justice focuses on the multiscalar interconnectedness of the human and more-than-human, and highlights the extent to which actions taken in one region, or even one community, impact the web of life across the planet. These concepts of planetary justice have become central to systems thinking in the 21st century (see Moore, 2015; Hornborg, 2020). Yet, if planetary justice is framed from above (for example, from an earth systems governance perspective), it risks becoming a general or universal idea of justice that perpetuates the centuries of those with hegemonic power speaking for one undifferentiated planet. Simpson and Pizarro Choy foreground the 'scalar tension' of planetary approaches to the climate crisis, warning that: 'Any top-down, singular global response to the climate crisis will most certainly mobilize existing global inequalities along axes of race, class, gender and citizenship in ways that insulate the beneficiaries of racial colonial capitalism from the most harmful impacts of the climate catastrophe while sacrificing others' (2023, p 2).

The approaches adopted by recent United Nations Climate Change Conferences (Conference of the Parties [COP] 27 and 28), far from reflecting an Earth system governance perspective, instead reflect the 'top-down' power and control exercised in the global economic core. This scalar tension is that the 'urgent need for coordinated action' risks forsaking local struggles and perpetuates 'epistemic violence against forms of Indigenous knowledge which offer crucial insights at this time of rapid ecological change' (Simpson and Pizarro Choy, 2023, pp 2–3). The challenge is how to forge a planetary approach that begins and attends to the specificity of local struggles.

This book's understanding of moving towards planetary justice draws on Gayatri Chakravorty Spivak's call for 'learning to learn from below, toward imagining planetarity' (2003, p 100). In a recent special issue of *Earth System Governance* dedicated to planetary justice, Hickey and Robeyns define 'planetary justice' as combining 'social justice' (justice 'in the wide array of social relations between persons, from cultural to political to economic relations') and 'environmental and ecological justice' (which foregrounds human relations with other species and ecologies) (2020, p 1). Twenty years before this definition, Spivak invoked the notion of planetarity in *Death of a Discipline* in response to globalization's 'imposition of the same system of exchange everywhere' (2003, p 72). Spivak foregrounded the planet's alterity: unknowable and invisible, in contrast to a globalization discourse claiming to make the 'global' visible and knowable (DeLoughrey, 2019, p 64).

With this planetary turn, then, humans become 'planetary subjects rather than global agents, planetary creatures rather than global entities' (Spivak, 2003, p 73). Spivak asks 'How many are we?' (2003, p 102), drawing attention to multiplicity and infinite difference. 'Planet thought', for Spivak, 'opens up to embrace an inexhaustible taxonomy of such names, including but not identical with the whole range of human universals' (2003, p 73). Gabrys et al (2022, p viii) extend and rework Spivak's focus on planetarity through 'planetarities' that aim to create a space for 'transdisciplinary and transnational movements' beyond the Western academy and institutional contexts. These plural understandings of the planet or planetary multiplicities resonate with Nigel Clark and Bronislaw Szerszynski's (2020, p 78) conceptualization of earthly multitudes or human and more-than-human compositions who participate in the 'immanent generation of difference' in response to the 'crisis' of climate change. The chapters in this book extend this participation, contributing to the planet's capacity to 'become otherwise'.

Understanding the human as a planetary subject (rather than a global subject) and a geological agent compels different ways of thinking about plural human and more-than-human futures (Spivak, 2003; Chakrabarty, 2021). Any questioning or visioning of 'futures', however, must also question 'whose futures and imaginaries' are being fought for, whether it is possible not to impose a vision of planetary justice on others and whether it is possible to build movements that are 'hospitable to a "pluriverse" of environmental imaginaries' (Collard et al, 2015, cited in Simpson and Pizarro Choy, 2023, p 3). These futures would affirm 'a deep relationality between the social and a more-than-ecological natural world' and unsettle the nature-culture binary as well as the white masculine gaze of planetary urbanization that dissolves difference through a 'totalizing analytic' (Derickson, 2018, pp 556, 557). Moving toward planetary justice is therefore a fundamentally radical political and ethical response to anthropogenic climate change.

The varied perspectives in this book offer fresh insights into the challenges of learning and living with climate change within and beyond academia. The assembled contributions seek to foster ongoing plurilogues about the politics and possibilities of creating different ways of living and relating to each other across and with this climate-changing planet. They refresh concepts of justice by being grounded in participatory action with frontline communities and reflections on praxis. Justice, in this book, is bottom-up, enacted, and emerges from interconnections, interdependence and ethical responsibility (Barnett, 2018). In keeping with Earth Unbound's plurilogical approach, the three parts in this book are woven together with short pieces that function as 'interstices'. Nishnaabeg philosopher, writer, poet and musician Leanne Simpson (in Maynard et al, 2021, p 142) explains the importance of interstitial places between knowledge systems, theories, struggles and lived experiences, which can be woven together in threads of vitality to recode and reorder the world. These interstices simultaneously stand apart and draw in alternative dimensions; interstices serve to bridge, bind and weave the book's parts together relationally. The interstices put the parts and chapters that precede them into conversation with each other and with those that follow.

Stories and studies – in three parts

Part I: Solidarity as responsibility, resurgence and regeneration

At a time when 'regeneration' and 'regenerative futures' have become buzzwords in climate change policy literature (Warden, 2021), the chapters in Part I open the book with fresh theoretical and empirical approaches and transdisciplinary debates on reinvigorating solidarity, collective responsibility, and justice for living and thriving otherwise. The chapters question and speak back to the violence of climate change through solidarities of collective reinvigoration, resurgence and responsibility. In Chapter 2, the reader's senses are awakened to the Law of the Land or living landscapes of the Martuwarra and Wardandi Noongar regions of Western Australia through a decolonial invitation that mobilizes solidarity and collective consciousness (see also Martuwarra et al, 2021). In Chapter 3, water colonialism, theft and trading in the Murray Darling Basin in Eastern Australia are exposed with the aim of refreshing traditional notions of environmental and ecological justice. Participatory research on freshwater availability, access and equity in the Indian Sundarbans of deltaic Bengal in Chapter 4 demonstrates how the expansion of a thriving water industry places the burdens of climate change on women and disadvantaged rural dwellers. In Chapter 5, sacred and kinetic energies from Gunditjmara and Wadawurrung Saltwater Country along the Southern Ocean, Australia mingle with *bhatir desh* ('Tide Country') in deltaic Bengal, India through experimental multisensory storying that creates decolonial crosscurrents of oceanic responsibility. Interstice 1 that follows reflects on the saturated strands of (insurgent/resurgent) solidarity that are woven through the chapters in this part.

Part II: Solidarity without borders

Part II is concerned with the questions: what are the dangers and promises of solidarity between privileged and marginalized groups in the struggle for planetary justice? What is the role of power and politics in the relationships between more and less powerful actors? Can common agendas be assumed and what are the risks of co-optation in myriad struggles against colonialism, capitalism and extractivism? These three chapters (Chapters 6, 7 and 8) focus on the tensions that emerge when equity and justice are lost or differentially understood in the framing of planetary justice by more powerful insiders or outsiders. Through the lens of Indigenous self-determination, Chapter 6 highlights differences between Northern and Southern activist groups in Australia and India in the imperatives, politics and narratives of resisting coal mining. Chapter 7 presents a detailed case study of the loss of localized Indigenous perspectives of development and wellbeing in the face of an alternative agenda driven by popular intellectuals. Chapter 8 highlights the limitations of a myopic climate activism for generating the structural and systemic change in the global economic core and considers how the 'fight' against climate change could be strengthened by engaging with a broader class struggle against capitalism and colonialism. All three chapters discuss possibilities for bridging divides in the interests of solidarity and planetary justice, including the imperative that the international working class and the land defenders lead the way. The second interstice by the Gesturing Towards Decolonial Futures Collective (GTDF) focuses on the Gifts of Failure. This is followed by Interstice 3 – Face to Face with the Supercyclone Amphan: Kolkata, 20 May, 2020 – by Sanjana Dutt, who uses vivid descriptions and heartfelt reflections to paint a poignant picture of resilience, solidarity and the profound impact of natural disasters on human lives.

Part III: Learning and living with climate change as situated solidarity

The chapters in Part III focus on learning and living with climate change, from situated experiences that mobilize solidarity but also foreground questions of power. Such solidarity unfolds across settings that are, and are not, conventionally understood to be 'places of learning' (Ellsworth, 2004). The authors consider the inseparable role of knowledge as a form of empowerment for action, and knowledge as an experiential, lived and embodied experience for responding directly to climate change. This part understands learning and living as interconnected: learning is both individual and collective (Kilgore, 1999), cognitive and corporeal (Ollis, 2012). 'Learning to live with climate change' (Verlie, 2022) necessarily involves unlearning extractivist modes of being and relating. Part III alternates between chapters situated within settings that are conventionally understood as 'educational' (for example, compulsory schooling) and sites of activist struggle that may not be immediately recognized as 'educational'.

Chapter 9 challenges colonial pedagogical approaches and argues for solidarity with Country and First Nations stakeholders to lead efforts toward planetary justice. In Chapter 10, the focus is on learning from lived experience in the Indian Sundarbans through knowledge co-production and situated adaptive practices as community adjustment 'tactics' to volatile ecologies. Chapter 11 reflects on collective involvement in a series of workshops on Education for Climate Justice, and develops a critical analysis of the relationship between the institutionalized disavowal of 'profane knowledge' and climate anxiety. Chapter 12 explores the experiences of the politics of solidarity for school-aged students involved in climate justice activism across intersectional differences. Part III concludes with an interstice that explores how these chapters exemplify activism, research, and pedagogy as forms of ecological praxis. This praxis might help address the void of responsibility for responding to planetary justice in mainstream education. A postscript poetically captures a dance of becoming with the world when the Earth is Undone.

Dancing with the Earth

Thinking with Tim Choy's (2021: 235) conceptual understanding of reckoning amid environmental violence, the stories and studies of solidarity in this book point 'to a direction forward' in multiscalar and unequal lifeworlds. Reorienting to the planetary from the grassroots up is urgent, given critiques of the limitations of normative climate justice frameworks that have run through academia, policy circles and grassroots activism over the last 20 years (Tschakert et al, 2020). When justice and solidarity are bottom-up, enacted and emerging from interconnections, interdependence and responsibility (Barnett, 2018), they become *otherwise,* mobilizing 'kinship, relationality and intersectional ethics' (Schmidt, 2022, p 2), and shaping ethical practices of struggle, resurgence and rebellion.

Rather than offering up any plans or utopian visions of the future, this book opens up spaces for considering multiple futures through wandering thoughts, actions and imaginations from Australia, India, Scotland and Canada. Such errantry, straying beyond the walls of the academy and scholarly norms, is urgent, given the often-deafening language of crisis that focus on the climate emergency, apocalypse and the sixth mass extinction. Rather than reinforce the dominance of a Western-centric worldview that emphasizes catastrophic thought and technological solutionism, we are inspired by the voices and embodied languages foregrounded in this edited collection. These voices highlight co-breathing (Choy, 2021) and a planetary consciousness 'that forces us to be born together again' with the Earth (Mbembe, 2022, n.p.). We hope this book might contribute to embodied praxes that unbind the Earth from injustices.

References

Akomolafe, B. (2021) 'We will dance with mountains' [Online course], Available from: https://www.bayoakomolafe.net/offerings [Accessed 10 June 2023].

Barnett, C. (2010) 'Geography and ethics: justice unbound', *Progress in Human Geography*, 35(2): 246–255.

Barnett, C. (2018) 'Geography and the priority of injustice', *Annals of the American Association of Geographers*, 108(2): 317–326.

Bawaka Country, Wright, S., Suchet-Pearson, S., Lloyd, K., Burarrwanga, L., Ganambarr, R., Ganambarr-Stubbs, M., Ganambarr, B., Maymuru, D. and Sweeney, J. (2016) 'Co-becoming Bawaka: towards a relational understanding of place/space', *Progress in Human Geography*, 40(4): 455–475, DOI: https://doi.org/10.1177/0309132515589437.

Bhambra, G.K., Gebrial, D. and Nişancıoğlu, K. (2018) 'Introduction: decolonising the University?' in G.K. Bhambra, D. Gebrial and K. Nişancıoğlu (eds) *Decolonising the University*, London: Pluto Press, pp 1–15.

Biermann, F., Dirth, E. and Kalfagianni, A. (2020) 'Planetary justice as a challenge for earth system governance', *Earth System Governance*, 6(2), DOI:10.1016/j.esg.2020.100085.

Bulkeley, H., Carmin, J., Castáa'n Broto, V., Edwards, G.A.S. and Fuller, S. (2013) 'Climate justice and global cities: mapping the emerging discourses', *Global Environmental Change*, 23(5): 914–925, DOI: 10.1016/j.gloenvcha.2013.05.010.

Byanyima, W. (2022) 'HIV or COVID-19, inequity is deadly', *Nature Human Behaviour*, 6(2): 176–176, Available from: https://www.nature.com/articles/s41562-022-01307-9.

Chakraborty, D. (2021) *The Climate of History in a Planetary Age*, Chicago: University of Chicago Press.

Choy, T. (2021) 'Externality, breathers, conspiracy: forms of atmospheric reckoning', in Papadopoulos, D., Puig de la Bellacasa, M. and Myers, N. (eds) *Reactivating Elements: Chemistry, Ecology, Practice*, Durham and London: Duke University Press, pp 231–256.

Clark, N. and Szerszynski, B. (2020) *Planetary Social Thought: The Anthropocene Challenge to the Social Sciences*, Chichester: Wiley.

Commonwealth of Australia (2020) 'Royal Commission into National Natural Disaster Arrangements Report', October, Available from: https://naturaldisaster.royalcommission.gov.au/system/files/2020-11/Royal%20Commission%20into%20National%20Natural%20Disaster%20Arrangements%20-%20Report%20%20%5Baccessible%5D.pdf [Accessed 28 March 2023].

Connolly, W. (2017) *Facing the Planetary: Entangled Humanism and the Politics of Swarming*, Durham, NC: Duke University Press.

Coulthard, G.S. (2015) 'The colonialism of the present' [Interview with Andrew Bard Epstein], *Jacobin*, Available from: https://jacobin.com/2015/01/indigenous-left-glen-coulthard-interview/ [Accessed 28 March 2023].

Crutzen, P.J. and Stoermer, E.F. (2000) 'The "Anthropocene"', *Global Change Newsletter*, 41: 17–18.

Curley, A., Gupta, P., Lookabaugh, L., Neubert, C. and Smith, S. (2022) 'Decolonisation is a political project: overcoming impasses between Indigenous sovereignty and abolition', *Antipode*, 54(4): 1043–1062.

Curnow, J. and Helferty, A. (2018) 'Contradictions of solidarity: Whiteness, settler coloniality, and the mainstream environmental movement', *Environment and Society: Advances in Research*, 9: 145–163.

DeLoughrey, E.M. (2019) *Allegories of the Anthropocene*, Durham, NC: Duke University Press.

Dembicki, G. (2020), 'A debate over racism has split one of the world's most famous climate groups', *Vice News* [online], 29 April, Available from: https://www.vice.com/en/article/jgey8k/a-debate-over-racism-has-split-one-of-the-worlds-most-famous-climate-groups [Accessed 17 July 2023].

Derickson, K. (2018) 'Masters of the universe', *Environment and Planning D: Society and Space*, 36(3): 556–562.

Ellsworth, E. (2004), *Places of Learning: Media, Architecture, Pedagogy*, New York: Routledge.

Escobar, A. (2020) *Pluriversal Politics: The Real and the Possible*, Durham, NC: Duke University Press.

Fitzgerald, R. (2023) 'Scientists send joint letter to NT government calling for ban on fracking in Beetaloo Basin', *ABC News*, 3 May, Available from: https://www.abc.net.au/news/2023-05-03/scientists-open-letter-nt-government-beetaloo-basin-fracking-ban/102291816 [Accessed 17 July 2023].

Fraser, N. (1997) *Justice Interruptus: Critical Reflections on the 'Postsocialist' Condition*, New York: Routledge.

Fraser, N. and Honneth, A. (2014) *Redistribution or Recognition? A Political-Philosophical Exchange*, London: Verso

Gabrys, J., Gray, R. and Sheikh, S. (2022) 'Planetarities: series foreword', in E. Glissant and P. Chamoiseau (eds) *Manifestos*, London: Goldsmiths Press, pp vii–viii.

Geoscience Australia (2022) 'Coal'. Australian Government, Available from: https://www.ga.gov.au/digital-publication/aecr2022/coal# [Accessed 17 July 2023].

Gómez-Barris, M. (2017) *The Extractive Zone: Social Ecologies and Decolonial Perspectives*, Durham, NC: Duke University Press.

Haraway, D.J. (2016) *Staying with the Trouble: Making Kin in the Chthulucene*, Durham, NC: Duke University Press.

Hickey, C. and Robeyns, I. (2020) 'Planetary justice: what can we learn from ethics and political philosophy?', *Earth System Governance*, 6: 100045.

Hornborg, A. (2020) 'The world-system and the earth system: struggles with the society/nature binary in world-system analysis and ecological Marxism', *Journal of World-Systems Research*, 26(2): 184–202.

Hornborg, A. and Martinez-Alier, J. (2016) 'Ecologically unequal exchange and ecological debt', *Journal of Political Ecology*, 23(1): 328–333.

Hunter, M. (2020) 'Race and the geographies of education: markets, White tone, and racial neoliberalism', *Annals of the American Association of Geographers*, 110 (4): 1224–1243, DOI: 10.1080/24694452.2019.1673144.

Jafry, T., Mikulewicz, M. and Helwig, K. (2018) 'Introduction: justice in the era of climate change', in T. Jafry (ed) *Routledge Handbook of Climate Justice*, New York: Routledge, pp 1–9.

Joseph-Salisbury, R. and Connelly, L. (2021) *Anti-racist Scholar-Activism*, Manchester: Manchester University Press.

Kilgore, D.W. (1999) 'Understanding learning in social movements: a theory of collective learning', *International Journal of Lifelong Education*, 18(3): 191–202.

Klein, N. (2022) '"Let's try something new": Naomi Klein calls for boycott of next COP climate summit', Available from: https://www.commondreams.org/news/2022/11/22/lets-try-something-new-naomi-klein-calls-boycott-next-cop-climate-summit [Accessed 22 June 2023].

Lahiri-Dutt, K. (2017) *Women at the Water's Edge. Lives of women in a climate changed world*, Available from: https://www.youtube.com/watch?v=2Bl2NkP9k9k [Accessed 3 June 2023].

Land, C. (2015) *Decolonizing Solidarity: Dilemmas and Directions for Supporters of Indigenous Struggles*, London: Zed Books.

Latour, B. (2018) *Down to Earth: Politics in the New Climatic Regime*, London: Polity Press.

Lobo, M., Bedford, L., Bellingham, R.A., Davies, K., Halafoff, A., Mayes, E., Sutton, B., Walsh, A.M., Stein, S. and Lucas, C. (2021) 'Earth unbound: climate change, activism and justice', *Educational Philosophy and Theory*, 1–20: DOI: 10.1080/00131857.2020.1866541.

Lobo, M. (2019) 'Affective ecologies: braiding urban worlds in Darwin, Australia', *Geoforum*, 106: 393–401.

Lobo, M. and Rodriguez D. (2022) 'Decolonising the university from the Antipodes: geographical thought and praxis', *Geographical Research*, 60(1): 40–45.

Mamdani, M. (2012) *Define and Rule: Native as Political Identity*, Cambridge, MA: Harvard University Press.

Maddrell, A., Lynn-Ee Ho, E. and Lobo, M. (2023) 'The multiple intensities of COVID-19 space-times', *Social and Cultural Geography*, 24(3–4): 385–390, DOI: 10.1080/14649365.2023.2177718.

Martuwarra RiverOfLife, Pelizzon, A., Poelina, A., Akhtar-Khavari, A., Clark, C., Laborde, S., Macpherson, E., O'Bryan, K., O'Donnell, E. and Page, J. (2021) 'Yoongoorrookoo: the emergence of ancestral personhood', *Griffith Law Review*, 30(3): 505–529.

Maynard, E., Simpson, L.B., Voegele, H. and Griffin, C. (2021) 'Every day we must get up and relearn the world: an interview with Robyn Maynard and Leanne Betasamosake Simpson', *Interfere*, 2: 140–165, Available from: https://hcommons.org/deposits/item/hc:51211/ [Accessed 5 June 2023].

Mbembe, A. (2022) 'How to develop a planetary consciousness: interview', *Noema*, Available from: https://www.noemamag.com/how-to-develop-a-planetary-consciousness/ [Accessed 5 June 2023].

Moore, J.W. (2015) *Capitalism in the Web of Life: Ecology and the Accumulation of Capital*, London: Verso.

Moore, J.W. (2019) 'Capitalocene and planetary justice: who is responsible for the climate crisis?', *Maize*, Available from: https://jasonwmoore.com/wp-content/uploads/2019/07/Moore-The-Capitalocene-and-Planetary-Justice-2019-Maize.pdf [Accessed 5 June 2023].

Nolan, K. (2019) *Aboriginal and Torres Strait Islander Timeline of Resistance*, Available from: https://s3-ap-southeast-2.amazonaws.com/originalpower/Original_Power_timeline_of_resistance_digital_version.pdf.

Ollis, T. (2012) *A Critical Pedagogy of Embodied Action: Learning to Become an activist*, Basingstoke: Palgrave Macmillan.

Parliament of Australia (2023a) Climate Change Amendment (Duty of Care and Intergenerational Climate Equity) Bill 2023. Available from: https://www.aph.gov.au/Parliamentary_Business/Bills_Legislation/Bills_Search_Results/Result?bId=s1385 [Accessed 10 December 2023].

Parliament of Australia (2023b) 'Senate Inquiry Report: oil and gas exploration and production in the Betaloo Basin', April, Available from: https://www.aph.gov.au/Parliamentary_Business/Committees/Senate/Environment_and_Communications/BeetalooBasin [Accessed 10 May 2023].

Parsons, M., Fisher, K. and Crease, R.P. (2021) *Decolonising Blue Spaces in the Anthropocene: Freshwater Management in Aotearoa New Zealand*, London: Palgrave Macmillan.

Puar, J.K. (2012) '"I would rather be a cyborg than a goddess": becoming-intersectional in assemblage theory', *philoSOPHIA*, 2(1): 49–66.

Pugliese, J. (2021) 'More-than-human lifeworlds, settler modalities of geno-ecocide and border questions', *Journal of Global Indigeneity*, 5(3): 1–34.

Roy, A. (2020) 'The pandemic is a portal,' *Financial Times*, 4 April.

Roshanravan, S. (2014) 'Motivating coalition: women of color and epistemic disobedience', *Hypatia*, 29(1): 41–58.

Ruddick, S. (2017) 'Rethinking the subject, reimagining worlds', *Dialogues in Human Geography*, 7(2): 119–139.

Shohat, E. (2001) 'Introduction', in E. Shohat (ed) *Talking Visions: Multicultural Feminism in a Transnational Age*, Cambridge, MA: MIT Press, pp 1–63.

Salleh, A. (2017) *Ecofeminism as Politics: Nature, Marx and the Postmodern*, London: Zed Books.

Satgar, V. (2018) 'The climate crisis and systemic alternatives', in Satgar, V. (ed) *The Climate Crisis: South African and Global Democratic Eco-Socialist Alternatives*, Johannesburg: Wits University Press, pp 1–28, Available from: https://www.jstor.org/stable/10.18772/22018020541.6?seq=1 [Accessed 10 December 2023].

Schlosberg, D. (2014) 'From environmental to climate justice: climate change and the discourse of environmental justice', *WIREs Climate Change*, DOI: 10.1002/wcc.275.

Schmidt, J.J. (2022) 'Geography and ethics I: placing injustice in the Anthropocene', *Progress in Human Geography*, DOI: 10.1177/03091325221097104.

Shiva, V. (1991) *The Violence of the Green Revolution: Third World Agriculture, Ecology and Politics*, London: Zed Books.

Simpson, M. and Pizarro Choy, A. (2023) 'Building decolonial climate justice movements: four tensions', *Dialogues in Human Geography*, DOI: 10.1177/20438206231174629.

Spivak, G.S. (2003) *Death of a Discipline*, New York: Columbia University Press.

Stengers, I. (2020) 'Foreword', in Green, L. (ed) *Rock, Water, Life: Ecology and Humanities for a Decolonial South Africa*, Durham: Duke University Press, pp xi–xv.

Sultana, F. (2021a) 'Climate change, COVID-19, and the co-production of injustices: a feminist reading of overlapping crises', *Social and Cultural Geography*, 22(4): 447–460.

Sultana, F. (2021b) 'Political ecology II: conjunctures, crises, and critical publics', *Progress in Human Geography*, 45(6): 1721–1730.

Sultana, F. (2022) 'The unbearable heaviness of climate coloniality', *Political Geography*, 99: 102638.

Táíwò, O.O. (2022) *Reconsidering Reparations*, Oxford: Oxford University Press.

Theriault, N., Leduc, T., Mitchell, A., Rubis, J.M. and Gaewohako, N.J. (2019) 'Living protocols: remaking worlds in the face of extinction', *Social and Cultural Geography*, 21(7): 893–908.

Tschakert, P., Schlosberg, D., Celermajer, D., Rickards, L., Winter, C., Thalers, M., Stewart-Harawira, M. and Verlie, B. (2020) 'Opinion: Multispecies justice: climate-just futures with, for and beyond humans', *WIREs Climate Change*, 1–10.

Tuck, E. and Yang, K.W. (2014) 'R-words: refusing research', in D. Paris and M.T. Winn (eds) *Humanizing Research: Decolonizing Qualitative Inquiry with Youth and Communities*, New York: Sage, pp 223–248.

Tuck, E. and Yang, K.W. (2018) 'Introduction: born under the rising sign of social justice', in E. Tuck and K.W. Yang (eds) *Toward What Justice? Describing Diverse Dreams of Justice in Education*, New York: Routledge, pp 1–17.

Unigwe, C. (2019) 'It's not just Greta Thunberg: why are we ignoring the developing world's inspiring activists?', *The Guardian*, 5 October.

United Nations Intergovernmental Panel on Climate Change (IPCC) (2023) *Sixth Assessment Report, AR6 Synthesis Report: Climate Change*, Available from: https://www.ipcc.ch/report/ar6/syr/ [Accessed 5 June 2023].

Vergès, F. (2019) *A Decolonial Feminism*, A.J. Bohrer (trans.), London: Pluto Press.

Verlie, B. (2022) *Learning to Live with Climate Change: From Anxiety to Transformation*, Abingdon: Routledge.

Walia, H. (2021) *Border and Rule: Global Migration, Capitalism, and the Rise of Racist Nationalism*, Chicago: Haymarket.

Warden, J. (2021) *From Sustaining to Thriving Together*, London: Royal Society for the Encouragement of Arts, Manufactures and Commerce (RSA).

Whyte, K.P. (2018) 'Indigenous science (fiction) for the Anthropocene: ancestral dystopias and fantasies of climate change crises', *Environment and Planning E: Nature and Space*, 1(1–2): 224–242.

Whyte, K.P. (2020a) 'Against crisis epistemology', in B. Hokowhitu, A. Moreton-Robinson, L. Tuhiwai-Smith, C. Andersen and S. Larkin (eds) *Routledge Handbook of Critical Indigenous Studies*, Abingdon: Routledge, pp 52–64.

Whyte, K.P. (2020b) 'Too late for indigenous climate justice: ecological and relational tipping points', *WIREs Climate Change*, DOI: 10.1002/wcc.603.

Wretched of the Earth Collective (2019) 'An open letter to Extinction Rebellion', *Red Pepper*, 3 May, https://www.redpepper.org.uk/an-open-letter-to-extinction-rebellion/ [Accessed 10 December 2023].

Yusoff, K. (2018) *A Billion Black Anthropocenes or None*, Minneapolis: University of Minnesota Press.

PART I

Solidarity as Responsibility, Resurgence and Regeneration

2

Waking up the Snake: Ancient Wisdom for Regeneration

Anne Poelina, Bill Webb, Sandra Wooltorton and Naomi Joy Godden

Opening

We are a collective of four Western Australians. Anne is a *Yimardoowarra marnin*: a Nyikina Indigenous woman who belongs to the Martuwarra Fitzroy River in the Kimberley (northwest) region. Bill is a Wardandi Noongar man who belongs to the southwest coastal region, where 'wardan' in the Noongar language refers to ocean and hinterland. He was one of the last of the 'fringe-dwellers' living in the Geographe Bay bush, which is now a tourist resort. Sandra is an Anglo-Celtic (Wadjela) woman whose family has been in Noongar *boodja* (Noongar Country)[1] for about eight generations, while Naomi is a Wadjela woman of European descent living on Wardandi *boodja*. We acknowledge Indigenous custodians and caregivers in our homelands, and in all parts of the world.[2]

Each of us is committed to biocultural stewardship, unity and collaboration, strengthening solidarity for climate action and planetary justice through ancient wisdom – First Law – for regeneration. By 'ancient wisdom', we mean the Law of the Land known to Elders, who hold these knowledges and teach them when the time is right for people to learn. Young Indigenous people are now taking these knowledges and making modern stories for planetary wellbeing (Poelina et al, 2022). We work to create eco social transformation in and through our community work, action research, writing, environmental education, public pedagogies, and by our collective wisdom and actions. Drawing attention to continuing colonization is part of this process.

In this chapter, we build on Indigenous environmental humanities literatures to frame the central argument with a deeply interconnected, Earth-centred nature of reality. Indigenous scholars like Lewis Williams

(2019, 2021) use onto-epistemological framings to bring to life the agency of a thinking, sentient, feeling Earth. We use Indigenous methodology and the conversational method (Kovach, 2019) to compose reflexive narratives through yarning – the *kapati* (cup of tea) method – from the Kimberley and Wardandi Noongar regions. *Kapati* is a method of story sharing within the safe space of conversation, which Ober (2017) notes has become a cultural norm in Indigenous communities and families. Our project used multiple cups of tea, sometimes with all four of us, sometimes in pairs (including mixed pairs) and often together with an audience to create stories and this chapter. Our narratives therefore take the form of small stories, poetry and autoethnography.

We recognise business-as-usual as continuing colonization, and illustrate the connection between decoloniality, climate justice and ecological health. The change we advocate is profoundly regenerative (Wahl, 2016; Poelina et al, 2022), emphasizing Indigenous knowledges of the vital nature of living places to reinvigorate life, love and living. We argue for collective consciousness as we come together for guardianship responsibility for Country; looking, listening, learning and practising while these life forces cleanse us together in a kindred spirit. This is transformative change of a type that we suggest is solutionary (rather than revolutionary). We seek solutions to right size the planet 'ground up', using ancient wisdom, which is living place-based Indigenous knowledge existing in a nonlinear time scale of past-presence-emergence. In our experience, relational onto-epistemologies inform local knowledges and practices that are vibrant, vital, practical, artistic/creative/dramatic, ceremonial and often invoke the energies of the sacred, the living Earth. These onto-epistemologies are ancient, drawing on multigenerational wisdom and experience.

We trust the world will wake up to First Law, the Law of the Land, which shows us how to perform an ethics of care, love and response-ability with more-than-human beings. By 'response-ability', we mean that humans can respond to the living, communicative, sentient world around us; to all our more-than-human kin with whom we live interwoven lives (Bawaka Country et al, 2019). Ancient wisdom transforms the banal into the beautiful, the significant and the sacred, a lifestyle of simplicity. We want to inspire an active narrative of hope for regeneration. Solidarity, in this sense, is responsibility, and response-ability is inclusive of our relations within our places; rivers, forests and other nonhuman co-inhabitants. These are living beings with agency. For this, we work towards socioecological reinvigoration *as* cultural resurgence.

Colonization as epistemic violence: decolonization as regeneration

Uncle Bill describes the invasion as the 'Taking Away'. He shares a Wardandi prophecy from thousands and thousands of years ago:

There will come a man to this land, and he will change it. He is not a sharer or carer. He won't leave a good footprint and he'll shape and change the land for himself. He'll never put back the things he's damaged. Overnight he'll turn the place into a desert, with species loss, deforestation, and degradation of rivers and waterways. He'll take away the dream of millennium-geared environmental scientists. He has no ambition to change his practice.

Uncle Bill explains that we must redress the viciousness of these people who removed our access to our land and did not allow it to become part of our earthly co-existence. There cannot be reconciliation and healing until they own up to the massacres and the indiscriminate killing and stealing of our people and spiritual places. Uncle Bill laments that the pristine waterway where Wardandi folks used to drink directly from the river, is now green and polluted. There is always an almighty dollar sign in the cloud. These stories need telling, and then we can move on as a nation.

Across Australia, Indigenous nations recognize a serpent as a creator spirit of rivers and waterways, which remains to protect water and First Law (Radcliffe-Brown, 1926). The metaphor of 'Waking up the Snake' is to wake up the consciousness of the people to restore the ancient wisdom of our places through an ethic of care and love (RiverOfLife et al, 2020). This Kimberley Indigenous metaphor restores obligations of stewardship and enables practices that increase awareness and consciousness that vital landscapes are animate, active, agential, inspirited and communicative. These obligations are First Law, the Law of the Land, and therefore apply to everyone without prejudice. In other words, once people walk, live, or do business on these Lands, whether borne of Country or as migrants, they carry these obligations of stewardship.

Indigenous nations across Australia know the sacred water snake by many names. In the Martuwarra Valley, Nyikina Indigenous peoples use Yoongoorookoo, an ungendered term (not he or she). In Noongar *Boodja* (Country) Noongar people know the sacred snake by several names, including Waugal. Uncle Bill shares a Waugal dreaming that connects the north and south of Australia:

> The big snake came from Flinders Ranges to Uluru and had two baby female snakes. One went to the north and one came to the south, to Wyadup on Wardandi *boodja*. The snake created all the waterways and rivers, ridges and mountains. She also had a baby and left it in the Yaragadee aquifer, and then she went back to the Flinders Ranges. Her baby snake remains in the Yaragadee to make sure that the beautiful Wardandi environment is cared for. The snake is our big overall connection and our belief of co-existence of all species.

We recognize Snake by its many names as integral to Australia's Earth dream, a long story of sentient worlds. Snake is dynamic, fluid, radiant and reconnective of places, beings, people, intuitions, spirits, archetypes and shadows. There is only relationship and connectedness in this worldview rather than separation.

In Australia, using the colonial paradigm as justification, corporations and governments still invade and colonize Indigenous communities and environments. Drafters of legislation frequently use colonial logics in ways that overlook Indigenous knowledge holders and their cultural interests. For these reasons, Australian laws, rules and policies such as the Aboriginal Cultural Heritage Act 2021 often do not serve the interests or the greater good of people or Country (Poelina et al, 2020). They do not relate to cultural-environmental interests where there is a mining or extraction claim. For instance, in 2020, Juukan Gorge, a 46,000-year-old record of human history in artistic form, was destroyed for iron ore mining[3] (Allam and Wahlquist, 2020).

Cultural stewardship as collective consciousness

Within this worldview we are sketching with Nyikina and Wardandi Noongar perspectives, stewardship of Country is enmeshed in a *relational* ethic of care and love. This is in the sense of caring for a more-than-human family within First Law (deriving from Country) over deep time, whereby intergenerational knowledge – that passed from one generation to another – is of great cultural significance. For stewardship of Country, knowledge systems are (always) cared for by the ancestors for present and future generations. This is known as a kincentric ecology, an ecology based on relationships, where environments are viable only where humans understand life around them as relation or kin (Salmon, 2000). It is a collective sense of self, where self is seen within a collective, and the more-than-human collective of people with Country is the greater interest rather than a narrow, egotistical sense of self. An example of this knowledge system can be seen in the Nyikina Calendar by Milgin et al (2020); a full colour chart of West Kimberley seasons. In this way, crises of sustainability are crises of relationship (Milgin et al, 2020). We are all beneficiaries of this relational, intergenerational knowledge through the care of precious places we now inhabit. By the same logic, we are ancestors for future generations of carers of Country.

Waking up the Snake for regeneration: collective action

Indigenous groups and alliances are currently reconstructing songlines along rivers and waterways in Australia, renewing awareness and prompting revitalization following years of colonial repression (for example, Gay'wu Group of Women, 2019; Mulligan et al, 2021). At the same time, local relationships and networks are extending outwards to link narratives across place, time and people, locally, regionally, nationally and internationally. This chapter records one of those reconnections, energizing regeneration through waking up consciousness of northwest-southwest Australian narratives through academic work.

As we conceptualize it, climate justice is climate care, that is, caring for all peoples, diversities, and everything – including ourselves – as part of our extended place-families or more-than-human kin networks through a radical transformation of oppressive systems of power (Godden et al, 2021; Whyte 2021). As Bawaka Country et al (2020) put it, we *are* our climates, in the sense that we co-become with local climate through knowing, doing and being. Acting in solidarity, together as one voice for climate justice, we recognize reinvigoration as the emergence of new stories of healthy connection for our planet at this time of conflict, famine, floods, fires and other emerging crises. From this point of view, environmental justice, climate justice and planetary justice co-arise with Indigenous justice.

Indigenous methodology

Kovach (2019) comments that those who use an Indigenous methodology need to place as much focus on that which connects the various elements as on the elements themselves. Thus, before we tell our stories, we declare the interconnected nature of our chapter with the Earth Unbound group with and for whom we developed this thinking. We participated in one planning conversation before a presentation to an Earth Unbound collaborative scholar group, who provided encouraging comments. Next, they supplied a transcription of our presentation with points of interest highlighted. With this, we collaboratively developed a proposal for submission to the editorial group, who offered substantial written conversational feedback for us. In this way, our ideas were already collaborative with the Earth Unbound group before we began to write.

Narratives

Anne: As an Indigenous leader, human and earth's rights advocate, my dream is the creation of peace, despite the violence that continues today. Yin Paradies writes that 'relatively scant attention [is being] paid to ongoing colonial presence/presents in which systemic, structural, physical, epistemic, and ontological violence continue to oppress, assimilate and eradicate Indigenous peoples' (2020, p 1). The pathway forward demands being brave and exploring the most appropriate way to support Indigenous peoples. We write and share stories of our colonial lived experiences to build reconciliation and a greater understanding of co-existence with our fellow Australians so that we can better understand how we, as human beings, can once again live in harmony with each other. When we refer to living in harmony with each other, we mean in ways that are mutually respectful of all more-than-human relatives, a term that includes people, nonhuman beings and Country – that is, in ways that are anti-racist, decolonial, underpinned by all forms of justice and that incorporate forms of development that use free, prior, informed and continuing consent. In other words, there will be no power-over, supremacy of any description or imposition of worldview, preference or any form of domination. The greatest challenge in the Kimberley is to create Peace with nature and Indigenous People, especially our young people.

Martuwarra Dream

Martuwarra Youth Council
Elders cry it can't be business as usual
'Sorry business' can't become the normal
Too many generations of lost lives
Kimberley holds highest suicide rates in the world
We Dream to hold them up and give them hope
Martuwarra Youth Council
Re-Generational Hubs are their life portal
Why is it so hard to get a hand up?
Culture camps Mum Annie's calling up
Go to Geegully, just a bough shed is all she wants
Aunty Annie does not get the call
But millions of dollars goes to the richest people of all
Public monies to fund the richest rich
A 'Feasibility study for an agriculture' they call it
Re-Generation Hubs – the Elders are dreaming this
Young people are standing, sharing, and dreaming
Why is it so hard to hear what our young people want?
We want to shift them from poverty, to work hard, to be free
Free to Dream and reach their full potential
We want them to experience being human
We want them to live in a Nation
Where Black lives matter
Where the Nation cares
Come on be brave and stand with us
Don't stand against us
We can help you to hear, feel and see your Country

Your home, you are now part of a new story
Let's make a stand, let's plan, let's send the Dream out
Invest in Martuwarra Youth Council.

Chinna and Poelina, in press

Poems such as this contain multiple stories to demonstrate risks, hopes and dreaming – all vital ingredients for building and sharing a spirit of goodwill from Indigenous Australians. Despite the violence and trauma, our intention is to reconcile with fellow Australians and with our Nation (Reconciliation Australia, 2016). It can't be business as usual; the planet and humanity cannot sustain it. The time is now to create peace with Indigenous people and with nature (Redvers et al, 2020).

Sandra: In the wet, firm beach sand there are footprints, bicycle tracks, snail trails, little balls of sand and tiny crab holes. A light mist over the ocean reveals a translucent glow. There is no wind. Beach beckons people towards her; irresistibly magnetic. Her gesture is palpable, stronger as we draw closer. Cable Beach, Broome, entices people, and thousands respond. She is alive, pulsing, vibrant and shimmering with memory. She tenders peacefulness. Beach loves life – song, dance, music, children laughing, youth playing, adults smiling, dogs running joyfully, crabs scurrying and birds playing in the waves.

Indigenous voices speak compellingly, re-storying and regenerating Country, cultures and histories. Thus, reciprocity with Country strengthens, while response-ability[4] grows. Indigenous authority, dignity and nationhood are re-emerging from colonial subjugation, inviting collectivism and union, and showing how to care for Country. Snake of living places re-emerges, waking up the consciousness of the people, interweaving Indigenous voices, sentient rivers, animate mountains, breathing forests, enlivened woodlands, storied places and solidarity renews wisdom in we who hear and respond. We feel the call of Indigenous Country as *longing to belong*. Country needs us to feel *in relation with* more-than-human worlds. This is the essence of decolonization; a journey for each and every one of us.

Indigenous people narrated Country for eternity. Then colonization, and silenced Indigenous voices took these stories to ground. Industrialists and agribusinesses now speed up resource extraction for the wealthy to get wealthier, while the poor get poorer and Country gets sliced, cut, mined and drained. Extractive colonization holds power over Indigenous people and Country, including we who stand together in solidarity. Complicity with colonization through our daily lives – in supermarkets, kitchens and energy systems – is destroying Country, stories and Snake, maintaining colonial disempowerment.

Waking up the Snake is to peel back layers in order to understand the impact of colonization, together with recognition and relearning of behaviours that reconnect. We *feel* sentient places when we care for Country through solidarity. There is hope because *you* are part of this story now.

Bill: My Wardandi mob has been around for over 65,000 years. Our Country was pristine, with a cultural and spiritual connection. We made sure everything was intact to pass on to our future generations, because rightfully it does not belong to us; it always belongs to the future. *Boodja* is the land and our heart. A lot of people are missing that real true connection to the land.

The Wardandi role in society is that we protect the spiritual pathways across land, of all spirits gathered. In my Country, the morning sun puts these spirits on a spirit trail across the land, moving into the caves comforted by *ngilgi* (the good sea spirit) and into the arms of Wardan (the great ocean spirit), to the ancient land of the setting sun (Kornup) and then dispersed in the cosmos reincarnated as a Jungara, a chosen time traveller. We all come from stardust, and the Jungara travels in space and time to the land to see if we are doing the right thing, looking after and caring for Land. He's not there to harm to you, he's there hoping that you recognize how important this place is right now in continuing ancient time. We can share our place, and these spiritual and always beautiful stories.

I can see that the children of today are confused about such questions as where their place is, how they're going to be in the future and which way they're going. They're searching. My dad had a vision that young people, whether they are Black, white or another ethnicity, will band together and make their future footprint as a unity of people. My role is building cultural awareness within people. I'm educating the kids, the future, how to band together to walk one journey. We must keep that old knowledge, that old culture and the spiritual connections with places, and these younger generations will all feel that collaborative harmony, culture and vibration energy consciousness. The future generations will be very responsible; they will take control through conscious learning from Indigenous peoples.

Naomi: Sometimes I look at my beautiful children and wonder, with profound grief, if I made a mistake by bringing them into this world. The climate science is undeniable – we are hurtling towards environmental, economic and societal collapse, and the growing collective demand for climate justice is criminally dismissed by governments of all persuasions. Climate change is a reckoning for colonialism, capitalism, Euro-patriarchy and white supremacy, and dismantling these oppressive systems of power is extremely difficult and painful. Hope can often feel fanciful and naïve, and many people fear for our children, all children, including children of all species. But, as Elders like Uncle Bill and Aunty Anne so generously and patiently teach us, generating collective consciousness through ancient wisdoms, particularly with our children, is our pathway forward.

I am a Wadjela woman who grew up, and is now growing up my own children, on Wardandi *boodja*. My family has deep love, attachment and spiritual connection to this place. With vulnerability and humility, I listen to Elders who share wisdom, stories and truth telling as we confront the insidiousness of ongoing colonization on this Country. Wardandi Elders have taught me that I am not separate from the interconnected system of *boodja*. I am *boodja*, and *boodja* is me. We are *boodja*, and *boodja* is us. Caring for the rivers, the forests, the more-than-human is caring for ourselves. This is not an intellectual process; it is a heart process, a spiritual process, an embodied process. As our consciousness expands and our egos diminish, we can conceptualize the more-than-human as our kin. Collectively, we are the Snake.

In the pursuit of climate justice, *boodja* and her Indigenous custodians are not victims, stakeholders or end users, but the central, interconnected system of sacred knowledge and wisdom from which all abounds. As the Snake awakens, humans in solidarity with *boodja* can (and must) dismantle oppressive colonial systems that cause climate injustice and work towards new (old) ways of collective knowing, being and doing that are grounded in Indigenous wisdoms.

Discussion

Readers of these narratives and perspectives recognize that concepts need a relational, imaginal, experiential, creative and connected sense of knowing, which is different from a Cartesian, Western, separated and individuated categorical sense in which business-as-usual thinking and knowing sit. This is not to say one way of knowing is more important; rather, it is to recognize their different contexts, languages, cultures, histories, purposes and outcomes. Recognizing this difference in worldview is to acknowledge that Indigenous knowledge holders have acted upon obligations arising from a cultural ethic of care, reciprocity and vitality for aeons. Waking up the Snake is liberating Country and the consciousness of the people to reconnect collectively as ecological families.

Each narrator writes with passion about defending and reconnecting with Country, strengthening Indigenous voices, and restoring agency to Country and more-than-human beings. We call for the liberation of the Snake of consciousness from repression, to voice the ancestral power of the continuing past once again for the sake of our children's children. Together we ask: why destroy Country and spirit along with our children's future and the stories that keep us strong for the continuing present? Like never before, we need to act in solidarity.

In his text *Sand Talk*, Tyson Yunkaporta proposes that if First Peoples' knowledges were respected 'as the most sustainable basis for living and being on this planet', Australian people would be able to move from 'sorry' to 'thankyou', and after a while, to 'please' (2019, p 267). These ancient wisdoms are a very generous offering as we (humans and more-than-humans) collectively grapple with the intersecting crises of climate change, environmental degradation and deepening social inequalities. Indeed, Robertson and Barrow (2020) explain that the Noongar people are very adept at surviving, and thriving, through climatic changes. They reference, for example, the continuation of Noongar peoples and cultural traditions through a 10,000-year drought. Likewise, the resilience of First Nations peoples in Australia persists through the brutal and continuing apocalypse of more than 240 years of colonial genocide.

Uncle Bill explains that technological solutions will not save us, and the least-impactful solution to climate change is the old way. He reinforces that if there were no harmful capitalist or colonial industries, there would be no social or environmental damage on this planet, and Indigenous people would still be living in harmony with Country. For our collective survival we must change how we think, joining the old ways and the new ways for future generations. This is explained in Poelina et al (2022), where they deploy Bird Rose's (2017) notion of the shimmer of life as ancestral power, a very different worldview from 'business as usual'. They explain:

> Shimmer comes with the pulse of season change, and the shininess and health of new growth, new generations. Shimmer might also be a pattern, or a dance, or a piece of art. It is the love story between species, such as the lure felt by the flying foxes to the angiosperms. It is the flourishing of life. (Poelina et al, 2022, p 5)

Uncle Bill attests that youth-led social movements such as the Seed Indigenous Youth Climate Network and School Strike 4 Climate (see Chapter 12) reflect his father's vision of young people uniting against a prevailing system that is obliterating biodiverse places, and certainly not doing the right thing for future generations. Of course, young people did not cause climate change and they are not responsible for resolving it (Godden et al, 2021). Yet, as a deeply conscious, progressive and benevolent generation, young people demonstrate our immense hope that the Snake is awakening and that collective consciousness is building. In contrast to the mainstream climate action movement, youth-led climate justice organizations in Australia actively respect Indigenous peoples' rights and knowledges, and work to protect Country, in their activism for radical structural change (see, for example, Seed, 2022).

Societal collapse due to climate cataclysm is a likely outcome of the lifeways of the West. Australians have the opportunity for a decolonizing process of rewilding, a returning to land and attending to neglected Country (Paradies, 2020) to systemically upend the capitalist addiction to the myth of human superiority over the planet. As Uncle Bill argues, we need to be people who are conscious of Indigenous cultures and how they successfully live their lives. Ancient beliefs will become a 'new' belief of creating collective consciousness for reinvigoration of First Law.

It has long been argued that Aboriginal knowledge holds solutions for planetary health, particularly climate disruption and species recovery. Writers such as Redvers et al (2020) and Milgin et al (2020) describe the intricacy and wisdom of interrelationship. Country – and River – holds kinships of inextricable rapport, which is the significance of this wisdom. Things – and all beings – have recently been seen as separate in the Cartesian worldview. Rather, the space between – the relationships of deep connection – holds the entangled consciousness of the Snake that is awakening. We acknowledge the joyfulness already emerging in Country, including the ceremonies and protocols of deep respect at conference openings. We conclude the discussion with one of the many living protocols that speak for us, offered by Theriault et al (2020), members of a transnational group of scholars, activists, artists and communities. One protocol is 'Our Teachers':

> We have been gifted with many teachers who continue to bring messages from the Creator on how to live according to the instructions of our ceremonies, songs, and speeches. We are fortunate to have so many young ones emerge from the earth who take our culture seriously. Our stories and songs guide us to stay in balance, and if we stray the Creator does not judge us but understands that we are on this difficult human journey. We say nya weh [thank you] for our teachers, and let it be that way in our minds. (Theriault et al, 2020, p 897)

Conclusion

In this chapter we make the point that colonization is commensurate with business as usual, and that we are each complicit in many ways due to the ubiquitous nature of interconnectedness. In other words, people are interrelated with nature for good or for bad; through business as usual with devastating consequences, and entirely differently, through Earth-kin with healing outcomes. Through four narratives from different perspectives, we show that the decolonial invitation through Indigenous onto-epistemologies and, in particular, ancient wisdom and First Law can lead to climate justice, the potential for life and ecological recovery. These are framings of hope that accompany the waking up of the Snake, a metaphor to reinvigorate the Indigenous consciousness of all people.

We advocate for Indigenous knowledges of the vital nature of living places, and solidarity with Indigenous voices. We argue for solidarity and collective consciousness as we come together for response-ability with Country and responsibility for Country; using creative, imaginal, artistic and experiential ways of knowing, being and doing (see also Chapter 9). The transformation we call for is cultural resurgence, which is solutionary. In this way, decoloniality, climate justice and ecological health allow for the possibility of a good life for young people. Ancient wisdom – the Law of the Land – shows the way to an ethics of care, love and response-ability with more-than-human beings. We hope you, the reader, will devise an active narrative of hope for regeneration with Indigenous teachers, and implement it.

Notes

[1] In Australia, Country (capital C) and Land (capital L) refer to a rich, living, spiritual notion of place that is inclusive of Indigenous earth-based laws and socioecological relationships, more-than-human beings and kin relationships. Australians tend to use Country rather than Land, whereas the reverse is true in the Americas. Also, this may be equivalent to 'Territory' in Latin American First Nations and Afro communities.

[2] This research is part of a consortium of the University of Notre Dame Australia, Edith Cowan University, Millennium Kids, the Pandanus Park Community, the Martuwarra Fitzroy River Council, Madjulla Inc, the WA Museum and the Water Corporation of Western Australia. It is financially supported (partially) by the following organizations: the Australian government through the Australian Research Council's *Linkage Projects* funding scheme (project LP210301390); the Water Corporation's Research and Development Program; the Millennium Kids Enviro Fund; and the WA Museum.

[3] In 2020, as part of its Pilbara mining operation, Rio Tinto blew up Juukan Gorge, a gallery of ancient arts with a sacred spring which had never stopped running – until now. There was an outcry in Australia and around the world about this act of terror. Rio Tinto responded by sacking a few senior staff members, apologizing and continuing its operations.

[4] 'Response-ability' refers to the ability to respond to Country or beyond (Bawaka Country et al, 2019).

References

Aboriginal Cultural Heritage Act (n.d.) [Online] Government of Western Australia, No. 27 of 2021 Stat. (2021). Available from: https://www.wa.gov.au/organisation/department-of-planning-lands-and-heritage/aboriginal-heritage-act-western-australia [Accessed 29 March 2024].

Allam, L. and Wahlquist, C. (2020) 'Gobsmacked: how to stop a disaster like Juukan Gorge happening again', *The Guardian* [online], 13 December, Available from: https://www.theguardian.com/australia-news/2020/dec/13/gobsmacked-how-to-stop-a-disaster-like-juukan-gorge-happening-again [Accessed 12 March 2024].

Bawaka Country, Suchet-Pearson, S., Wright, S., Lloyd, K., Tofa, M., Sweeney, J., […] and Maymuru, D. (2019) 'Goŋ Gurtha: enacting response-abilities as situated co-becoming', *Environment and Planning D: Society and Space*, 37(4): 682–702.

Bawaka Country, Wright, S., Suchet-Pearson, S., Lloyd, K., Burarrwanga, L., Ganambarr, R., […] and Maymuru, D. (2020) 'Gathering of the clouds: attending to Indigenous understandings of time and climate through songspirals', *Geoforum*, 108: 295–304.

Bird Rose, D. (2017) 'Shimmer: When all you love is being trashed', in A. Tsing, H. Swanson, E. Gan and N. Bubandt (eds) *Arts of Living on a Damaged Planet*, Minneapolis: University of Minnesota Press, pp 51–63.

Chinna, N. and Poelina, A. (in press) *Tossed up by the Beak of a Cormorant: Poems of Martuwarra Fitzroy River*, Fremantle, WA: Fremantle Press.

Gay'wu Group of Women (2019) *Songspirals: Sharing Women's Wisdom of Country through Songlines*, Sydney: Allen & Unwin.

Ghosh, A. (2021) *The Nutmeg's Curse: Parables for a Planet in Crisis*. Chicago: University of Chicago Press.

Godden, N.J., Farrant, B.M., Yallup Farrant, J., Heyink, E., Carot Collins, E., Burgemeister, B., […]. and Cooper, T. (2021) 'Climate change, activism, and supporting the mental health of children and young people: perspectives from Western Australia', *Journal of Paediatrics and Child Health*, 57(11): 1759–1764.

Kovach, M. (2019) 'Conversational method in Indigenous research', *First Peoples Child and Family Review*, 14(1): 123–136, Available from: https://fpcfr.com/index.php/FPCFR/article/view/376/308 [Accessed 12 March 2024].

Milgin, A., Nardea, L. and Grey, H. (2020) 'Birr Nganka inmany jada Warloongaryi yoonoo Woonyoomboo-ni: the cycle of life, story and law in Nyikina Country, as shared by Woonyoomboo. A Nyikina seasonal calendar', Available from: https://nesplandscapes.edu.au/publications/nyikina-seasonal-calendar/ [Accessed 12 March 2024].

Milgin, A., Nardea, L., Grey, H., Laborde, S. and Jackson, S. (2020) 'Sustainability crises are crises of relationship: learning from Nyikina ecology and ethics', *People and Nature*, 2020(2): 1210–1222.

Mulligan, R., Yamera, M., Warbie, J., Street, M., Poelina, A., Mulligan, E. and Coles-Smith, M. (2021) 'The Marlaloo songline', *AIATSIS Collection*, Canberra: AIATSIS, Madjulla Inc.

Ober, R. (2017) 'Kapati time: storytelling as a data collection method in Indigenous research', *Learning Communities: International Journal of Learning in Social Contexts*, 22: 8–15, DOI: 10.18793/LCJ2017.22.02.

Paradies, Y. (2020) 'Unsettling truths: modernity, (de-)coloniality and Indigenous futures', *Postcolonial Studies*, 23(4): 438–456.

Poelina, A., Brueckner, M. and McDuffie, M. (2020) 'For the greater good? Questioning the social licence of extractive-led development in Western Australia's Martuwarra Fitzroy River region', *The Extractive Industries and Society*, 8(3), DOI: 10.1016/j.exis.2020.10.010.

Poelina, A., Wooltorton, S., Blaise, M., Aniere, C.L., Horwitz, P., White, P.J. and Muecke, S. (2022) 'Regeneration time: ancient wisdom for planetary wellbeing', *Australian Journal of Environmental Education*, 1–18, DOI: https://doi.org/10.1017/aee.2021.34.

Radcliffe-Brown, A.R. (1926) 'The Rainbow-Serpent Myth of Australia', *Journal of the Royal Anthropological Institute of Great Britain and Ireland*, 56: 19–25.

Reconciliation Australia (2016) 'The state of reconciliation in Australia – a summary: our history, our story, our future', 22 February, Available from: https://www.reconciliation.org.au/publication/2016-state-of-reconciliation-in-australia/ [Accessed 12 March 2024].

Redvers, N., Poelina, A., Schultz, C., Kobei, D.M., Githaiga, C., Perdrisat, M., [...] and Blondin, B. (2020) 'Indigenous natural and First Law in planetary health', *Challenges*, 11(2): 1–12, DOI: 10.3390/challe11020029.

RiverOfLife, M., Poelina, A., Bagnall, D. and Lim, M. (2020) 'Recognizing the Martuwarra's First Law right to life as a living ancestral being', *Transnational Environmental Law*, 9(3): 541–568.

Robertson, F. and Barrow, J. (2020) 'A review of Nyoongar responses to severe climate change and the threat of epidemic disease: lessons from their past', *International Journal of Critical Indigenous Studies*, 13(2): 123–138.

Salmon, E. (2000) 'Kincentric ecology: Indigenous perceptions of the human-nature relationship', *Ecological Applications*, 10(5): 1327–1332.

Seed (2022) *Seed Indigenous Youth Climate Network*, Available from: https://www.seedmob.org.au/ [Accessed 12 March 2024].

Theriault, N., Leduc, T., Mitchell, A., Rubis, J.M. and Jacobs Gaehowako, N. (2020) 'Living protocols: remaking worlds in the face of extinction', *Social and Cultural Geography*, 21(7): 893–908.

Wahl, D.C. (2016) *Designing Regenerative Cultures*, Axminister: Triarchy Press.

Whyte, K. (2021) 'Time as Kinship', in J. Cohen and S. Foote (eds) *The Cambridge Companion to Environmental Humanities*, Cambridge: Cambridge University Press, pp 39–54.

Williams, L. (2019) 'Reshaping colonial subjectivities through the language of the land', *Ecopsychology*, 11(3): 174–181, DOI: https://doi.org/10.1089/eco.2018.0077.

Williams, L. (2021) *Indigenous Intergenerational Resilience: Confronting Cultural and Ecological Crisis*, Abingdon: Routledge.

Yunkaporta, T. (2019) *Sand Talk: How Indigenous Thinking Can Save the World*, Melbourne: Text Publishing.

3

Farmers as Allies Towards Ecological Justice: Lessons from Water Markets, Colonialism and Theft in Australia's Murray-Darling Basin

Alexander Baird

Introduction

Fresh water is the lifeblood for all human and nonhuman species. It is fundamental to our human wellbeing, global ecosystems, all flora and fauna, and supports all human industry, from tourism to mining and from agriculture to aquaculture. However, the uneven commodification and distribution of freshwater resources around the globe has led to a situation where water scarcity is on the rise, especially under increasing pressures from drought and climate change (see Chapter 4). This situation compounds difficulties faced by farmers, including in Australia, to acquire, protect and balance freshwater for its instrumental and intrinsic value. This chapter explores the experiences and perceptions of a purposive sample of local dairy, horticulture and viniculture farmers in the Darling, Lachlan, Murray and Murrumbidgee River systems in the Murray-Darling Basin (MDB) in Australia. Interviews were conducted, as a part of my PhD study, to explore farmers' experiences and perceptions about how freshwater is exploited, traded, managed and stolen in the MDB (Baird, 2023). Local farmers were selected for my research because, compared to corporate agribusinesses, farmers maintain an intimate and extensive knowledge of local ecosystems (Hickell, 2020).

This study draws from the field of green criminology, which has aligned itself with a critical harms-focused approach that emphasizes dire but perfectly legal crimes and harms, many commissioned by state and corporate actors. Thus, green criminology is concerned with the political, economic, sociological and cultural basis of crimes and harms, and the concomitant need for environmental protection, justice and sustainability that exposes the nature of environmental offending and victimization (Brisman and South, 2013). While concerned with the political and economic infrastructures that give rise to environmental injustices, green criminology also involves comparing, investigating and analysing environmental and ecological justice discourses that promote both the perpetration of environmental harm and crime and its prevention (White, 2021). Specifically, environmental justice is an anthropocentric concept, whereby humankind lies at the centre of all existence. From this perspective, the natural environment should be transformed and protected to accommodate the daily necessities for human beings and future human generations. Environmental justice discourses have been criticized for reinforcing the belief that the natural environment, including all plants and animals, can be commodified, traded and exploited as 'resources' for human benefit (White, 2018; Walters, 2019).

In contrast to environmental justice discourses, ecological justice refers to the relationship and interaction of human beings with the rest of the natural world and includes concerns relating to the health of all components and inhabitants of the biosphere (White, 2021). From this perspective, ecosystems are at the centre of the ontology, and human beings are only one component of the complex system that has intrinsic value beyond sustaining human life. Ecological justice encourages human beings to recognize both the instrumental and the intrinsic value of all living and nonliving components and inhabitants of the biosphere and to prioritize their rights to existence. This includes the rights of nonhuman species to live free from torture, abuse and destruction of habitat (Walters, 2019). This chapter foregrounds the need to shift from an environmental to an ecological justice lens that aligns with Indigenous-led activism to recognize the instrumental and intrinsic value of freshwater and legal rights of rivers in the MDB.

Australia's Murray-Darling Basin

Aboriginal peoples have occupied the landscape of the MDB as traditional custodians for tens of thousands of years. Today, more than 40 different Aboriginal tribes hold a unique connection to Country in the MDB, which not only represents their origins and ties to ancient geological structures and landscapes, but also comprises their spiritual and ontological connection to land, nature and past ancestors. The riverine plains and valleys have also shaped the cultural landscape of Aboriginal peoples and retain vital significance within their social life and varied customary economies

(Humphries, 2007; Ross and Ward, 2009; Weir, 2010). However, European colonization saw the arrival of early settlers who brought foreign plants and animals, and an ambition to transform the land, while shrouded in the myths of 'terra' and 'aqua nullius'; land and water belonging to no one (Marshall, 2017, p 224). European settlers set about establishing agrarian societies modelled on their European homeland, which have created much of the social fabric of Australia today. The efforts of early settlers to control and 'civilize' the Australian landscape for agrarian development established their supposed moral proprietorship over natural resources, but forcibly removed and dispossessed Aboriginal peoples of their sociocultural connections to land and water (Humphries, 2007; Downey and Clune, 2020). The riparian landscape and inland waterways were physically altered and transformed through extensive river constructions to support human settlements, irrigation, grazing and mining. This process has continued from the 19th century to today, where water is now governed and managed through a complex system of regulation and a water market that prioritizes corporate agribusiness (Connell, 2011; Wheeler et al, 2014).

The MDB is Australia's largest and most regulated freshwater river system and the twentieth-largest river catchment in the world. It provides vital freshwater resources for invaluable and fragile ecosystems and wildlife across Australia, and is relied on by irrigators and local communities (Wei et al, 2011; Douglas et al, 2016; Grafton and Wheeler, 2018). Today, the MDB is administratively split and the 'water resources' are shared between the State jurisdictions of Queensland, New South Wales (NSW), the Australian Capital Territory (ACT), Victoria and South Australia. Commonly referred to as the nation's 'food bowl', the MDB is home to more than two million people, generates 40 per cent of Australia's agricultural production and accounts for 66 per cent of Australia's agricultural water use. While fresh water used was once managed as a local commons, it is now traded through complex licence arrangements tenured under individual legal ownership, and even bottled and delivered to consumers through private corporate networks for a profit – all facilitated by state bureaucracies and regulatory arrangements (Johnson et al, 2017a, 2017b; Brisman et al, 2018; Baird and Walters, 2020). Freshwater is a profitable commodity to be traded on markets, where drought and water scarcity only serve to increase its monetary value (Eman, 2016; see also Chapter 4). Not surprisingly, freshwater is increasingly referred to as 'liquid gold' or 'blue gold' (Barlow and Clarke, 2017). Water markets, including those in Australia, have enabled corporate monopolies and governments to capitalize and commodify what is a fundamental human right for the pursuit of profit, with devastating environmental and human consequences (Johnson et al, 2017a; Brisman et al, 2020). In what follows, I will examine the commodification of water within the contexts of environmental harm, racism and injustice.

The commodification of water in Australia

Scholars and activists drawing on principles of environmental justice recognize that in current farming practices, freshwater resources are prioritized and protected in order to further economic development over more ecologically based considerations (White, 2021). In 2004, the Australian government separated land and water rights across the MDB, which enabled farmers to trade water licences and supplies between catchments and districts. The water market allows farmers to supplement their water supply in the short and the long term, earn income from selling water rights and expand agricultural production (ACCC, 2021). However, over the past two decades, the separation of land and water rights has seen the emergence of new nonlandholder market participants that treat the water markets similar to stock markets to increase, diversify and liquidate investment assets for personal and corporate financial gain (ACCC, 2021). While this has assisted some irrigators to develop new products to manage future risks of water scarcity (Seidl et al, 2020), investments in the water markets, and the rules that govern them, were designed by all levels of Australian government to benefit corporate agribusiness. The operation of the water markets in Australia is increasingly recognized as unfair to the critical water needs of local farmers who produce food and fibre for the Australian population. One farmer highlighted the way in which the water market mechanism in Australia disadvantages smaller farmers:

> The divorcing of water from land and the allowance there for water to be traded, you have created a situation where water can be traded to the highest value crop. That creates winners and losers, and the perception of some that the system is unfair ... It's a perception of unfairness in that allocation is owned by the big corporates, the overseas hedge funds and water traders, it is not owned by dairy farmers or small operators. (Baird, 2023, p 282)

The water market architecture that defines water supply and metering policies in the MDB varies between Queensland, NSW, Victoria, South Australia and the ACT. This, complexity, combined with lax rules and limited oversight for trading conduct – alongside labyrinthine water governance and regulatory arrangements already in place to manage water – creates opportunities for investors to exploit market flaws, distort trade information and impound water for private gain (ACCC, 2021). Another farmer in my research explained that this is 'essentially a redistribution of wealth from the common person to irrigators, corporate irrigators, bankers, institutional investors etc ... and the common person, our environment, our communities are paying the price' (Baird, 2023, p 283). Apart from the lack of a level playing field between local farmers, communities and corporate agribusiness in the water market, the market arrangements that prioritize freshwater for corporate benefit can be understood through the lens of environmental racism.

Environmental racism and water injustice

Environmental racism is concerned with the unequal distribution of natural resources and examines how the most vulnerable populations are least able to benefit from natural endowments, while being routinely exposed to social hazards and environmental harms (Walters, 2019). Indigenous water injustice is a form of environmental racism that exists in the context of what Hartwig et al (2021) refer to as 'water colonialism'. This entails past and present acts, institutions and discourses of water management that exclude Indigenous populations (Hartwig et al, 2021). Contemporary agricultural norms and practices that appear as economically neutral and beneficial in fact perpetuate past colonial injustices that underpin and exacerbate water colonialism (Hartwig et al, 2021). Past injustices, such as displacement, dispossession and environmental racism, increase the difficulty for Aboriginal people in the MDB to acquire water allocations and compete in the water markets, and undermine the recognition and legitimacy of Aboriginal water rights that are seen to stand outside of, or as irrelevant to, the formal economy (Hartwig et al, 2018, 2021). In the MDB, for example, Aboriginal communities represent 6.5 per cent of the total population, but hold only 0.17 per cent of available surface water holdings (Hartwig and Jackson, 2020). Indeed, water colonialism to this day dispossesses Aboriginal communities of their land and water, and compounds difficulties faced by Aboriginal people in terms of accumulating wealth and securing water rights, power and recognition under settler colonial systems of water administration (Hartwig et al, 2021).

A culture of overextraction, mismanagement and malfeasance, based on decades of exploitation and harmful water policy since European settlement, has given rise to the state-facilitated theft of water. Collusive arrangements between state-corporate actors have enabled opportunities for unmetered floodplain water to be harvested and impounded without a licence (Baird, 2023). Floodplain harvesting is one of the more egregious water challenges and manifestations of water theft in the MDB, and involves the diversion and impoundment of water across floodplains into privately owned on-farm dams to support large-scale agricultural and economic development in the Northern MDB, specifically NSW (Wheeler et al, 2020). The impacts of floodplain harvesting extend to erode community wellbeing and cohesion through social and economic decay, and are most pronounced on vulnerable communities downstream, such as Menindee in the Southern Basin (Baird, 2023). The decline of inland flows from upstream has resulted in poor water quality for drinking and recreational usage, leading to physical and mental health conditions to the point where local communities, such as the remote town of Wilcannia – where almost two thirds of the population are Aboriginal (ABS, 2021) – have been forced to source water from Australian cities (Maloney et al, 2020). One farmer explained:

> Do you mean justice or injustice? Well, when you have towns of a thousand people with no water … whether that is injustice or just a violation of human rights … it is a violation of Australia's general attitude that we are a developed nation contributing to all of the rules of the world and holding ourselves up there as having a high standard of living, and we are expecting those people up there to drink bottled water and shower in mud. (Baird, 2023, p 292)

The fundamental human rights to water prescribed by the United Nations General Assembly (Resolution 64/292) recognizes that 'clean drinking water and sanitation are essential to the realisation of all human rights' (cited in Johnson et al, 2017a, p 1). Australia is recognized as a global leader in water management due to its science-driven policy, innovative water markets and forward planning (Global Water Institute, 2017). However, water colonialism and theft in the MDB undermine human rights to water by decreasing access and supply for those who cannot afford it and increasing profits for those who can (Baird, 2023). The Australian government's legitimacy as a leader in global water management also falls well short in comparison to other nations, such as Italy, Bolivia and Slovenia, which have established a permanent moratorium on water privatization (Eman and Meško, 2021).

Local farmers in Australia manage significant risks, including highly variable and unpredictable changes in climate, drought and volatile commodity prices that impact the availability of water needed for agricultural production. Nowadays, with the advent of climate change, farmers in the MDB contend with higher temperatures and lower winter rainfall (Department of Agriculture, Water and the Environment, 2022). While environmental justice aims to safeguard agricultural production and human rights to water as a *resource*, it is increasingly clear that an anthropocentric approach to water management is unsustainable in the long term. The environmental justice perspective which focuses on human rights has led to a theft of environmental flows from the ecosystems in the MDB. However, in discussions with farmers, the environmental impacts of water theft were often couched in terms of intergenerational justice and the rights of human children. One interviewee highlighted the centrality of rivers to the wellbeing of future generations:

> It is terrifying for us to think what our next generation going to do, because we can see how much damage has been done to the Country, and we have only got a little bit of river Country in proportion to

what we farm, but it is such an important productive Country for us ... we can't do it if there is no river because the river is the core of what keeps that Country healthy. (Baird, 2023, p 299)

In this chapter, environmental justice foregrounds traditional notions of water justice by, for example, recognizing and prioritizing the instrumental value of freshwater resources for the human economy and society over ecological considerations. We have seen how in the MDB, freshwater resources have been commodified and privatized to further economic and agricultural development since European settlement and how this has resulted in inequity in water access and colonialism. This began with the establishment of white settler agrarian societies that were economically dependent on agricultural production and farming families to consolidate social ties and trade within the rural community (Botterill, 2009). In doing so, the Australian landscape and ecology have been severely impacted and continue to be harmed. Over time, the corporate takeover of agricultural land and water is shifting the balance against local farmers, who, in turn are beginning to question their impact on the health of the river system.

I turn next to consider an ecological justice perspective to highlight the centrality of healthy river ecosystems in the web of life now and for future generations. I outline how the recognition by local farmers of the impacts of water commodification and theft on ecosystems is turning the tide away from understanding water as a 'resource' to be bought and sold to the highest bidder. The next section draws some hopeful insights with local farmers who, faced with indisputable evidence that the river systems on which they rely are in serious peril, are determined to pursue ecologically just alternatives to water management, while safeguarding its instrumental value.

Solidarity as necessity

Between 2016 and 2020, at the height of the drought, water theft through floodplain harvesting led to a significant decline of inland flows and generated serious environmental harm along the Darling River and into the Menindee Lakes. Farmers interviewed for the research discussed how massive river red gums that were hundreds of years old were dying due to a lack of flowing water and how riverbeds were drying out and wetlands were returning to a terrestrial state through desertification. They noted that there were tremendous losses in terms of the numbers and species diversity of wildlife, including fish, freshwater mussels and migratory birds (Baird, 2023). The predictable lack of flows due to floodplain harvesting, excessive water extraction and reduction in the quality of water through the drought were the direct result of regulatory failure and water theft, and can therefore be considered ecocide. In just one event in 2019, millions of native fish, including native and endangered cod populations, suffocated and died within stratified, algae-infested pools of water in the Lower Darling River and Menindee Lakes systems (Australian Academy of Sciences, 2019). As one interviewee explained:

> The impacts we were told about and that we saw ourselves was just the smell of death, particularly when we went out to Menindee, Wilcannia, anywhere. The river system just stunk, the little bit of remnant water that was there was filthy, green, there were dead animals everywhere to Menindee ... dead kangaroos, dead emus, dead everything, an absence of life, an absence of animals. (Baird, 2023, p 183)

Four years later, the Darling River has experienced another mass kill in 2023 with millions of fish suffocating and unable to escape from hypoxic waters. While this kill has been linked to recent floods in NSW, 'the fundamental reason the fish of the Darling keep dying is because there is not enough water allowed to flow' and too much water is diverted for upstream irrigation (Kingsford, 2023, p 23).

Wetland ecosystems in the MDB are also under threat. These wetlands support high levels of biodiversity, providing vital breeding grounds and habitats for numerous species of wildlife, and contribute to regional economies by providing environments for fishing, grazing and tourism (EPA, 2021). However, the ecological health and conditions of wetlands in NSW and the abundance and diversity of wildlife that rely on them for support are declining fast. This is largely due to the reduction in water availability 'caused by altered flows from water extraction and the building of dams, levees and diversion structures, as well as by climate change, exacerbated by extreme weather events like heatwaves and droughts' (EPA, 2021, p 87). Only 20 per cent of floodplain wetlands targeted for environmental flows received an effective flood between 2014 and 2019 (Chen et al, 2020; Slezak, 2020). The mismanagement of environmental flows has diminished the magnitude, duration and extent of flooding required to maintain and improve the conditions of wetlands (Doody et al, 2015; Chen et al, 2020).

Of relevance to this chapter, an ecological justice perspective extends to the rights of rivers to flow freely without harm and disruption from human activities (White, 2018). In ecological justice, water bodies are recognized for their intrinsic value and are perceived as independent beings with agency that is autonomous of human control. Thus, harmful acts committed against freshwater would not only be conceived as harming those that depend on water's life-giving capabilities, but would also be seen as victimizing and violating the rights and liberties of the water body itself. Such harmful acts include water pollution and altering and preventing the flow of water. Ecological justice perspectives align with and draw on Indigenous knowledges that foreground spiritual and ontological entanglement of the more-than-human as a construct of their personal and cultural identity (White, 2021). Ecocentrism asserts that legal rights ought to be assigned to nonhuman entities that reflect their core traits and characteristics and, in the case of a river, this would involve the river flowing freely without harm and disruption from human activities (White, 2018).

More than 100 organizations from around the globe have so far endorsed the Universal Declaration on the RightsofRivers (2022). Environmental activists, scholars, policy makers and Indigenous groups are mobilizing to assign the status and rights of legal personhood to rivers of ecological and cultural significance. This sees rivers as equal to humans in the eyes of the law (Clark et al, 2018; White, 2018). For example, in 2017, the Indigenous Māori people of Whanganui, namely the Whanganui Iwi, and the New Zealand government established the Te Awa Tupua (Whanganui River Claims Settlement) Act that recognized the legal personhood of the Whanganui River. This saw the Whanganui River as a singular, living entity indivisible from the kinship and spiritual connections of its people to the river, with the same rights as a legal person that can be judicially enforced by two appointed guardians (Blankestijn and Martin, 2018; Argyrou and Hummels, 2019). In 2016, Traditional Owners from six Martuwarra Indigenous Nations created the Fitzroy River Declaration which represents an agreed expression of First Law and the rights to life for the sacred Martuwarra RiverOfLife (Martuwarra RiverOfLife et al, 2020) in Western Australia. For Traditional Owners, 'the Martuwarra is an integrated and whole living system from source to sea, head to tail. The Martuwarra is therefore a single living entity with an equal right to life' (Martuwarra RiverOfLife et al, 2020, p 548). In 2018, the Martuwarra Nations established the Martuwarra Fitzroy River Council, which manages and protects the Martuwarra River and its sovereignty from further human interference.

In my research, one interviewee personified 'the river and the environment as the biggest persons impacted' (Baird, 2023, p 301) on the ecocentric continuum in the MDB. This contrasts with the commodification, atomization and segmentation of rivers through water allocations that reflect economic and political priorities. By excluding considerations of intrinsic value which view a river as a single free-flowing entity, its ability to acquire equal legal recognition to people is undermined (Clark et al, 2018). This also extends to nearby ecosystems that support the river system, including the iconic river red gum trees that provide nesting opportunities for waterbirds in the tree canopy and habitats for snakes and small animals on the ground. But water resource development for human benefit has reduced the magnitude of floodwaters that red gum trees need to survive, causing their health and condition to deteriorate over time and impacting the entire river ecosystem (Doody et al, 2015).

Farmers as allies?

Ecological citizenship implies that humans must prioritize universal human interests and rights, such as water and food security (environmental justice), and govern their actions to ensure the ecological wellbeing of the biosphere (ecological justice) (White, 2021). This transcends the legal notions of rights and citizenship to consider humans as part of planetary ecosystems, breaking down the binary construct of 'human' and 'nature'. Some farmers in my research portrayed 'ecological citizenship' that was embedded in a deep respect and sense of protection for various inland waterways in the MDB. This, most fundamentally, involved recognizing and promoting the need to prevent water theft from the riverine ecosystems. For example, one farmer asserted that the rivers should flow freely 'because there is a whole range of ecological, social, and economic values that are being degraded or destroyed by large water extractions associated with irrigation' (Baird, 2023, p 305) upstream in NSW. Local farmers and residents have long argued that the social and economic wellbeing of rural communities are intrinsically linked to the health of the Darling River (Maloney et al, 2020). An ongoing flow of water is needed to manage the river system and deliver water to local farmers and communities, for which the environment receives the benefit of conveyance water lost through evaporation, transpiration and seepage into the soil (ACCC, 2021).

However, apart from meeting human needs, the ecosystem needs were recognized by many of the farmers in my interviews. One farmer suggested that water for upstream irrigation should be moved to and prioritized for downstream irrigation because the 'virtue of the delivery of water to that person or water holder [downstream] is actually bringing something that creates environmental benefits' (Baird, 2023, p 305). This enables the river system upstream to flow freely for its own sake and protects the intrinsic right of the river to flow, providing water for wildlife and surrounding ecosystems along the way before being extracted for its instrumental value downstream. Another interviewee explained that some local farmers also use their own water and land to reduce the decline of wetlands downstream in the MDB. For example, one farmer said: 'you have got ... people that will quarter off part of their land for the environment, will undertake environmental water with their own water and their own property and will voluntarily let go of some of their water for downstream wetlands ... threatened birds and frogs' (Baird, 2023, p 306).

One farmer also discussed the increasing interest and engagement of local farmers in regenerative farming principles and practices. Regenerative farming principles and practices aim to increase biodiversity, enrich soil health, improve watersheds and enhance ecological services (Burns, 2020). This farmer intentionally designates a portion of their water allocation for the surrounding environment and maintains a deep respect for water through their regenerative farming operation. During an interview, they explained how the process reflects a deep commitment to ecological justice:

> I do not see myself, personally the way that I farm, being in competition with the environment, I see myself as a part of the environment ... I farm regeneratively. I see the importance of the environment in delivering and properly managing waterways for environmental services as critical in delivering good quality water. I also see my operation as contributing environmental services. (Baird, 2023, p 307)

Regenerative farming encourages farmers to shift away from ecologically degrading large-scale commercial farming methods to holistic thinking that finds new ways to work with nature by mimicking natural systems, generating soil fertility, and placing emphasis on the role of water and land instead of technology (Burns, 2020). This idea of regeneration not only reflects but also encourages farmers to engage in a process called 'degrowth'; a planned downscaling of energy and resource use that reduces capitalist expansion and brings the economy back into balance with the natural environment in a safe, just and equitable manner (Hickell, 2020). Hickell notes that: 'When you treat a farm like an ecosystem instead of a factory, you begin a relationship with the land that is inimical to the short-term extractivist logic of agribusiness' (2020, p 250).

Contemporary water management practices and markets, predicated on traditional notions of environmental justice since European colonization, are ill-equipped to deal with the novel challenges of climate change and the radical changes to agricultural practice this necessitates. Uncertainties about water privatization, colonialism and theft, coupled with drought and floods exacerbated by river system mismanagement, have compounded difficulties faced by farmers to receive and allocate freshwater resources. Therefore, understanding the role of those farmers as allies in countering water injustice in all its forms provides an optimistic perspective of change. It is important to recognize that increasing numbers of farmers are pursuing water management alternatives that are based on ecological justice and citizenship. They reject the binary logic of 'human' and 'nature', and recognize both the instrumental and intrinsic value of freshwater and promote degrowth in the process. Doing so has strengthened their social, economic and spiritual connections with the ecosystems in which they see themselves as entangled.

This chapter has explored how local farmers – as witnesses to the impacts of water theft and mismanagement in the MDB – are becoming aware of new ways of managing freshwater supplies in the MDB through strategies committed to the goals of both environmental and ecological justice. While most farmers I spoke to in my research were concerned with protecting the instrumental value and usage of water to safeguard economic and agricultural production in the MDB, even they had begun to question the social and intergenerational impacts. Others had moved from seeing water justice through the lens of environmental justice to ecological justice and had found ways to act in solidarity with ecosystems to manage freshwater supplies through ecological citizenship and regeneration that balances fresh water's instrumental and intrinsic value.

References

Argyrou, A. and Hummels, H. (2019) 'Legal personality and economic livelihood of the Whanganui River: a call for community entrepreneurship', *Water International*, 44(6–7): 752–768.

Australian Academy of Sciences (2019) 'Investigation of the causes of mass fish kills in the Menindee Region NSW over the summer of 2018–2019', *Australian Academy of Sciences*, 18 February, Available from: https://www.science.org.au/files/userfiles/support/reports-and-plans/2019/academy-science-report-mass-fish-kills-digital.pdf [Accessed 5 May 2022].

Australian Bureau of Statistics (ABS) (2021) 'Wilcannia [2021 Census Aboriginal and/or Torres Strait Islander people QuickStats]', *Australian Bureau of Statistics*, Commonwealth of Australia, Available from: https://abs.gov.au/census/find-census-data/quickstats/2021/ILOC10300506 [Accessed 30 January 2023].

Australian Competition and Consumer Commission (ACCC) (2021) *Murray-Darling Basin Water Markets Inquiry: Final Report*, Available from https://www.accc.gov.au/about-us/publications/murray-darling-basin-water-markets-inquiry-final-report [Accessed 30 January 2023].

Baird, A. (2023) 'Drain the state: exploring water theft in Australia's Murray-Darling Basin', PhD thesis, Deakin University.

Baird, A. and Walters, R. (2020) 'Water theft through the ages: insights for green criminology', *Critical Criminology*, 10: 1–18.

Barlow, M. and Clarke, T. (2017) *Blue Gold: The Battle against Corporate Theft of the World's Water*, London: Earthscan Publications/Routledge.

Blankestijn, W. and Martin, A. (2018) 'Testing the (legal) waters: interpreting the political representation of a river with rights in New Zealand' Master's thesis, Faculty of Natural Resources and Agricultural Resources, Swedish University of Agricultural Sciences.

Botterill, L.C. (2009) 'The role of agrarian sentiment in Australian rural policy', in F. Merlan and D. Raftery (eds) *Tracking Rural Change: Community, Policy and Technology in Australia, New Zealand and Europe*, Canberra: Australian National University Press, pp 59–78.

Brisman, A. and South, N. (2013) 'A green-cultural criminology: an exploratory outline', *Crime, Media, Culture*, 9(2): 115–135.

Brisman, A., McClanahan, B., South, N. and Walters, R. (2018) *Water, Crime and Security in the Twenty-First Century: Too Dirty, Too Little, Too Much*, London: Palgrave Macmillan.

Brisman, A., McClanahan, B., South, N. and Walters, R. (2020) 'The politics of water rights: scarcity, sovereignty and security', in K. Eman, G. Mesko, L. Segato and M. Migliorini (eds) *Water, Governance and Crime*, New York: Springer, pp 17–29.

Burns, E.A. (2020) 'Thinking sociologically about regenerative agriculture', *New Zealand Sociology*, 35(2): 189–213.

Chen, Y., Colloff, M.J., Lukasiewicz, A. and Pittock, J. (2020) 'A trickle, not a flood: environmental watering in the Murray-Darling Basin, Australia', *Marine and Freshwater Research*, 72(5): 601–619.

Clark, C., Emmanouil, N., Page, J. and Pelizzon, A. (2018) 'Can you hear the rivers sing: legal personhood, ontology, and the nitty-gritty of governance', *Ecology LQ*, 45: 787–844.

Connell, D. (2011) 'Water reform and the federal system in the Murray-Darling Basin', *Water Resources Management*, 25(15): 3993–4003.

Department of Agriculture, Water and the Environment (2022) 'Snapshot of Australian agriculture', Available from: https://www.awe.gov.au/abares/products/insights/snapshot-of-australian-agriculture-2022#around-72-of-agricultural-output-is-exported [Accessed 28 June 2022].

Doody, T.M., Colloff, M.J., Davies, M., Koul, V., Benyon, R.G. and Nagler, P.L. (2015) 'Quantifying water requirements of riparian river red gum (Eucalyptus camaldulensis) in the Murray-Darling Basin, Australia: implications for the management of environmental flows', *Ecohydrology*, 8(8): 1471–1487.

Douglas, E.M., Wheeler, S.A., Smith, D.J., Overton, I.C., Gray, S.A., Doody, T.M. and Crossman, N.D. (2016) 'Using mental-modelling to explore how irrigators in the Murray-Darling Basin make water-use decisions', *Journal of Hydrology: Regional Studies*, 6: 1–12.

Downey, H. and Clune, T. (2020) 'How does the discourse surrounding the Murray-Darling Basin manage the concept of entitlement to water?', *Critical Social Policy*, 40(1): 108–129.

Eman, K. (2016) 'Water crimes: a contemporary (security) issue', *CRIMEN-časopis za krivične nauke*, 1: 44–57.

Eman, K. and Meško, G. (2021) 'Water crimes and governance: the Slovenian perspective', *International Criminology*, 1: 208–219.

Environment Protection Authority (EPA) (2021) *NSW State of the Environment Report*, Available from: https://www.soe.epa.nsw.gov.au/sites/default/files/2022-02/21p3448-nsw-state-of-the-environment-2021_0.pdf [Accessed 2 March 2022].

Global Water Institute (2017) 'Challenges, successes and priorities in Australia's water management', Available from: https://www.globalwaterinstitute.unsw.edu.au/news/challenges-successes-and-priorities-in-australias-water-management [Accessed 30 October 2021].

Grafton, R.Q. and Wheeler, S.A. (2018) 'Economics of water recovery in the Murray-Darling Basin, Australia', *Annual Review of Resource Economics*, 10: 487–510.

Hartwig, D. and Jackson, S. (2020) 'The status of Aboriginal water holdings in the Murray-Darling Basin', Available from: https://research-repository.griffith.edu.au/bitstream/handle/10072/400302/Jackson455456Published.pdf?sequence=5&isAllowed=y [Accessed 1 November 2021].

Hartwig, D., Jackson, S. and Osbourne, N. (2018) 'Recognition of Barkandji water rights in Australian settler-colonial water regimes', *Resources*, 7(1): 1–12.

Hartwig, L.D., Jackson, S., Markham, F. and Osborne, N. (2021) 'Water colonialism and Indigenous water justice in south-eastern Australia', *International Journal of Water Resources Development*, 38(1): 30–63.

Hickell, J. (2020) *Less Is More: How Degrowth Will Save the World*, London: William Heinemann.

Humphries, P. (2007) 'Historical Indigenous use of aquatic resources in Australia's Murray-Darling Basin, and its implications for river management', *Ecological Management & Restoration*, 8(2): 106–113.

Johnson, H., South, N. and Walters, R. (2017a) 'The commodification and exploitation of fresh water: property, human rights and green criminology', *International Journal of Law, Crime and Justice*, 44: 146–162.

Johnson, H., South, N. and Walters, R. (2017b) 'Eco crime and freshwater', in M. Hall, T. Wyatt, N. South, A. Nurse and G. Potter (eds) *Greening Criminology in the 21st Century: Contemporary Debates and Future Directions in the Study of Environmental Harm*, Abingdon: Routledge, pp 133–146.

Kingsford, R. (2023) 'How did millions of fish die gasping in the Darling – after three years of rain?', *The Conversation*, 20 March, Available from: https://theconversation.com/how-did-millions-of-fish-die-gasping-in-the-darling-after-three-years-of-rain-202125 [Accessed 28 March 2023].

Maloney, M., Boehringer, G., MacCarrick, G., Satija, M., Graham, M. and Williams, R. (2020) 'The 2019 Inquiry into the Health of the Barka/Darling River and Menindee Lakes', Available from: https://tribunal.org.au/wp-content/uploads/2020/10/2019CitizensInquiry_BarkaDarlingMenindee-1.pdf [Accessed 30 September 2020].

Marshall, V. (2017) 'Overturning aqua nullius: pathways to national law reform', in R. Levy, M. O'Brien, S. Rice, P. Ridge and M. Thornton (eds) *New Directions for Law in Australia*, Canberra: Australian National University Press, pp 221–230.

Martuwarra RiverOfLife., Poelina, A., Bagnall, D. and Lim, M. (2020) 'Recognising the Martuwarra's First Law right to life as a living ancestral being', *Transnational Environmental Law*, 9(3): 541–568.

RightsofRivers (2022) 'Universal Declaration on the Rights of Rivers', Available from: https://www.rightsofrivers.org/#declaration [Accessed 28 March 2023].

Ross, S. and Ward, N. (2009) 'Mapping Indigenous peoples' contemporary relationships to country: the way forward for Native Title and natural resources management', *Reform*, 93: 37–40.

Seidl, C., Wheeler, S.A. and Zuo, A. (2020) 'Treating water markets like stock markets: key water market reform lessons in the Murray-Darling Basin', *Journal of Hydrology*, 581: 1124399.

Slezak, M. (2020) 'Water from Murray-Darling Basin plan not being delivered to wetlands, Australian-first report finds', *ABC News* [online], 17 November, Available from: https://www.abc.net.au/news/2020-11-17/murray-darling-missing-water-in-floodplains/12887342 [Accessed 10 November 2021].

Walters, R. (2019) 'Green justice', in P. Carlin and F.L. Ayres (eds) *Justice Alternatives*, Abingdon: Routledge, pp 42–59.

Wei, Y., Langford, J., Willett, I.R., Barlow, S. and Lyle, C. (2011) 'Is irrigated agriculture in the Murray-Darling Basin well prepared to deal with reductions in water availability?', *Global Environmental Change*, 21(3): 906–916.

Weir, J.K. (2010) 'Cultural flows in Murray River country', *Australian Humanities Review*, 48: 131–142.

Wheeler, S., Loch, A., Zuo, A. and Bjornlund, H. (2014) 'Reviewing the adoption and impact of water markets in the Murray-Darling Basin, Australia', *Journal of Hydrology*, 518: 28–41.

Wheeler, S.A., Carmody, E., Grafton, R.Q., Kingsford, R.T. and Zuo, A. (2020) 'The rebound effect on water extraction from subsidising irrigation infrastructure in Australia', *Resources, Conservation and Recycling*, 159: 104755.

White, R. (2018) 'Ecocentrism and criminal justice', *Theoretical Criminology*, 22(3): 342–362.

White, R. (2021) *Theorising Green Criminology: Selected Essays*, Abingdon: Routledge.

4

Freshwater Access, Equity and Empowerment in the Indian Sundarban Region

Anwesha Haldar, Kalyan Rudra and Lakshminarayan Satpati

Thinking about freshwater

The Sundarban ecosystem situated in the lowermost part of the Ganga-Brahmaputra-Meghna-Delta region (GBM) in India and Bangladesh has been witnessing a severe freshwater crisis (Figure 4.1). The crisis is caused by a lack of storage facilities, the presence of deep inaccessible aquifers, as well as cyclonic storms and saltwater incursions exacerbated by climate change (Rudra, 2019). Freshwater scarcity, water pollution and contamination have led to serious concerns among the local people and the authorities. In addition, a thriving water business has developed, creating sharp divisions among the residents whose access to potable water depends on their economic status. In these circumstances, we argue that extensive rainwater harvesting is crucial to the sustainable development of water resources and climate justice in the region. But moving towards climate justice requires responding to development through proper planning, implementation of water policies and maintenance of water infrastructures.

To understand water availability, access and equity, this chapter is informed by over a decade of participatory grassroots research in the Indian Sundarban Region.[1] We draw on recent policy and media reports (in Bengali and English), participant observation, interviews, focus groups and surveys with selected households in 19 Community Development Blocks (CDBs) of the North and South 24 Parganas districts of West Bengal. These CDBs include Mathurapur-II, Kultali, Sagar, Patharpratima, Gosaba and Basanti, among others (Figure 4.2). During the field visits in March 2022, we explored different aspects of water access, quality and use, through 88 structured interviews and group discussions at household and community levels with men and women aged 18 to 65 years. We also held discussions with local farmers, traders, shopkeepers, businessmen, political leaders and stakeholders from non-governmental/voluntary organizations. Guiding our discussions were questions on issues pertaining to water and sanitation needs, especially among girls and women. We asked participants how they accessed water for domestic use and how they were affected by poor water quality. This chapter therefore highlights key findings from our decade-long research, focusing on the water crisis in the Indian part of the Sundarban tracts, where climate change and the decay of freshwater creeks is affecting humans, plants and animals.

Climate change and water crisis in the Indian Sundarban Region

With the increasing trend in disasters due to anthropogenic climate change in the mangrove delta of the Indian Sundarban Region (ISR) with its fragile network of tidal channels, earth system governance and planetary stewardship is crucial for healthy water systems. Such planetary stewardship would enhance human wellbeing and the sustainability of ecosystems that nourish fish, aquatic plants, animals and other species through the formulation and integration of actions across scales, from the local to global (Seitzinger et al, 2012; Vajpayee, 2023). The 'Only One Earth' slogan on the fiftieth anniversary of World Environment Day 2022 is a reminder that access to clean water and sanitation remains a crucial Sustainable Development Goal (SDG 6; see United Nations, 2022). Despite a large number of policies and programmes on sustainability and climate change mitigation at the global and local levels, there are few benefits for residents, in particular women, in the ISR.

The low-lying ISR, with 104 islands covering an area of 6,000 square kilometres, is affected by high-magnitude tropical cyclones that originate in the Bay of Bengal (BoB) in the Indian Ocean (Figure 4.1). In recent decades, storms have increased in intensity, and there has been a rise in global warming-induced sea surface temperature, as well as high tidal fluctuations (Haldar, 2015; Mitra et al, 2018). In addition, the tidal channels of the Sundarban with freshwater sources are decaying through histories and geographies of human occupation and land use practices. With the growing population in the post-independence era (1947 onwards) and rapid urbanization, particularly since the 1970s, settlements have grown on lands that were under forested areas. In the ISR, tidal creeks are frequently flooded by storm surges, and wetlands (*beels*, *jola* and *baors*) as well as tanks excavated for collecting surface water have been filled up. Therefore, the surface storage of fresh water from precipitation and the volume of water

Figure 4.1: Location map of the Indian Sundarban Region

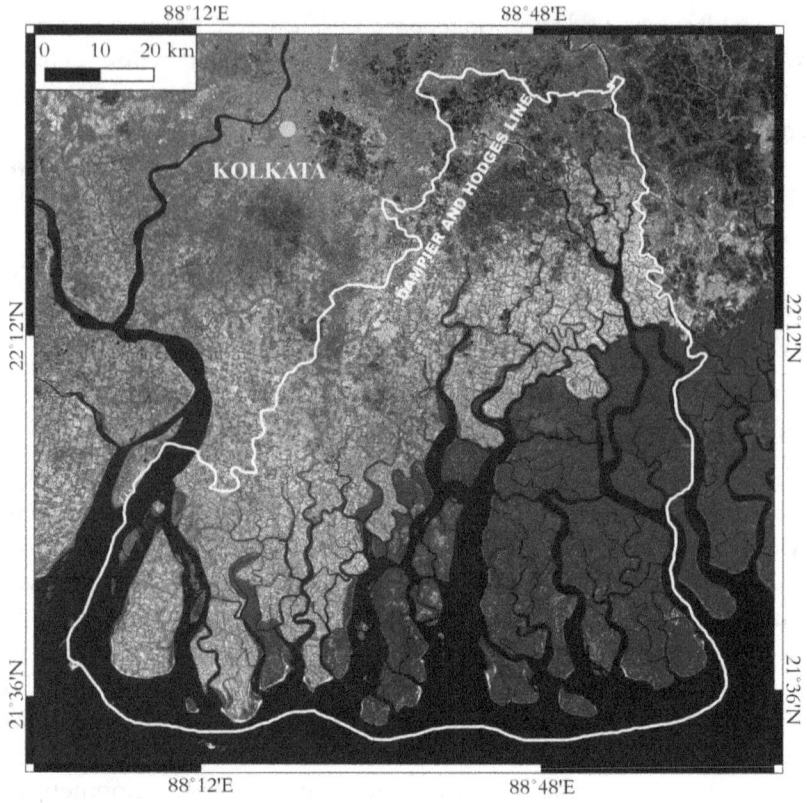

Source: Prepared by the authors based on Census of India administrative boundaries (Census, 2011)

in near-surface aquifers have drastically reduced in recent decades. These changes, together with groundwater sources prone to arsenic and fluoride contamination and the ingression of seawater into aquifers, have impacted communities in Sagar Island, Patharpratima, Gosaba, Jharkahli, Pathankhali and many other parts of the Sundarban (Das et al, 2021). The outcome has caused serious damage to the natural and built environment, including the breaching of embankments, the washing away of settlements and limited access to safe drinking water.

Safe drinking water is a fundamental human need and a basic human right (Rockström and Gordon, 2001; do Amaral et al, 2023). In India, however, 330 million people (approximately 26 per cent of the total population) face drinking water scarcity and agricultural distress (UNICEF, 2021). Therefore, even though water is found in abundance in the estuarine tracts of the Sundarban at the head of the BoB, as a freshwater resource it is scarce (Hazra et al, 2019). The Sundarban region has therefore been identified as

Figure 4.2: Community Development Block map of the Indian Sundarban Region

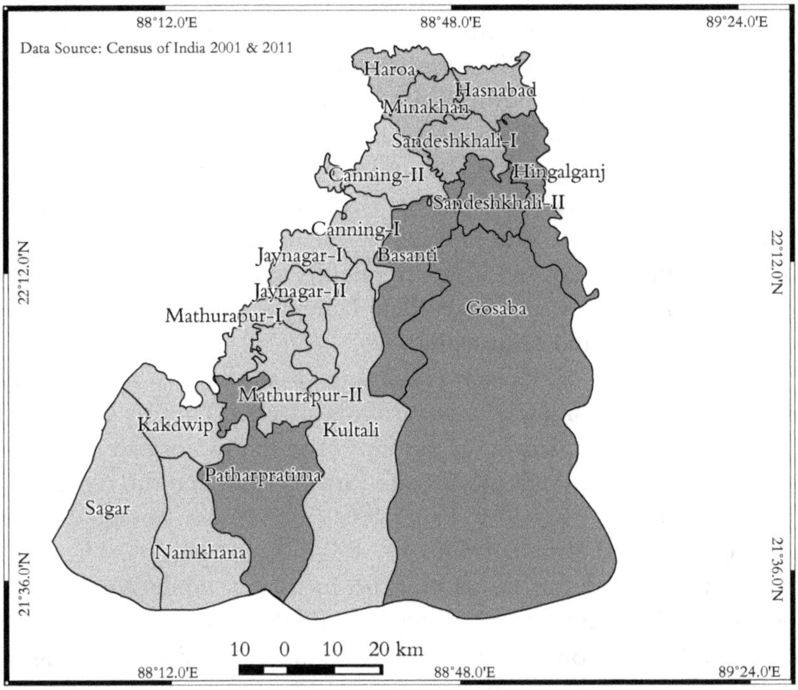

Source: Prepared by the authors based on Census of India administrative boundaries (Census, 2011)

a 'Critically Vulnerable Coastal Area' (Ministry of Environment, Forest and Climate Change [Government of India], 2019). In areas between 200 to 500 metres from the coast, withdrawal through ordinary wells is permitted, ensuring that deep freshwater is used for drinking purposes only (Bhadra et al, 2020). So far, the United Nations Sustainable Development Goal 6, Clean Water and Sanitation (United Nations, 2022), the Water, Sanitation and Hygiene (WASH) programme of the United Nations International Children's Emergency Fund (UNICEF), the 'Swajal Mission' of the Government of India (Ministry of Jal Sakti, 2019) and the West Bengal Drinking Water Sector Development Programmes (WBDWSDP), have had limited impacts on the provision of clean drinking water in the ISR; local people face severe shortages (Ministry of Drinking Water and Sanitation, 2018).

The freshwater crisis affects the livelihoods of the growing population who depend on agriculture, animal husbandry, pisciculture, horticulture and tourism. The average annual rainfall in the North and South 24 Parganas Districts may be adequate for monsoon agriculture, but during the drier seasons, failure to conserve excess rainwater has often led to agricultural droughts. In Gosaba CDB, most farmers who use deep tube wells and shallow pumps underline that salt encrustation on the soil takes about three monsoon seasons to clear (Haldar and Debnath, 2014). In addition, surface runoff from monsoonal downpours and cyclonic events leading to floods have disastrous impacts in the premonsoon and postmonsoon seasons on sources of water consumption. Tube wells are immersed in saline flood waters, open defecation flows onto the roads and settlements, the drain water meets surface water sources, and effluents from ancillary industries of the adjacent Kolkata Metropolitan Area (KMA) flow into this runoff. Given this decline in the water quality and usability of surface water, water security is compromised. In villages frequently affected by cyclonic storms, local healthcare workers in Sagar Island lamented child mortalities due to waterborne diseases and poor sanitation.

The water quality of shallow aquifers in places like Sagar, Namkhana, Patharpratima and Basanti CDBs are seriously affected. The clay layers impede quick infiltration to replenish the amount of groundwater extracted, as the recharge points of the deep aquifers are further away, and the subterranean groundwater flow is very slow. Therefore, groundwater storage cannot be adequately sustained if the current rate of withdrawal persists. The aquifers up to 60 metres in depth are all brackish in nature, while 70–160 metres below ground level, it is mostly saline. Despite this lack of freshwater access in these CDBs, the concentration of population is rapidly increasing. The total domestic demand for water was around 16 million cubic metres (mcm) in the ISR recorded in the 2011 census year. With an annual population growth rate of around 1.5 per cent, the water demand in the domestic sector is estimated to reach 44 million cubic meters in 2050 (Halder et al, 2021b). In addition, the availability of sweet water in the deeper aquifers, at 300 to 600 metres below ground level, is not only costly for the installation of deep tube wells but also unsustainable due to the low natural recharge rate and saline seawater intrusions (Bhadra et al, 2018).

Unequal community access to water

Groundwater-based piped water supply covers only 32 per cent of households of the Sundarban area (World Bank, 2014), located mostly on the mainland. The overall situation suggests that the entire ISR is heading towards a severe water crisis in the near future. This has led to a thriving water business, given the popularity of – and, indeed, reliance on – packaged water over the last two decades. As water has become a profitable yet essential commodity, it has also created a sharp division among people in terms of access to safe potable water according to their economic status. Waterborne diseases, arsenic contamination and other pollutants in the water affect the most vulnerable.

In the ISR, water demand in the domestic and agriculture sectors is met by deep tube wells. For example, rice cultivation in nonmonsoon seasons is only possible due to shallow submersible pumps and deep tube wells, which are few in number and affect the storage and availability of sweet water in the deep aquifers of the region. Moreover, only the larger-scale farmers can install them because installation, operation and maintenance costs are high and therefore unaffordable for the small and marginal farmer. Very often, saline water is deliberately inducted into shallow ponds for shrimp farming, and thus sweet water sources are further reduced and deteriorated. More recently, export-oriented commercial shrimp farming requires mixing of large quantities of sweet water, usually obtained by pumping out groundwater. In some cases, this not only pollutes and degrades the water environment of the region, but also creates water conflicts among communities, as the potable water sources among vulnerable people decline due to the lowering of the freshwater aquifer.

A water business in the agriculture sector has developed in the ISR, with small and marginal farmers being forced to purchase groundwater to irrigate their cultivated plots. Impoverished small and marginal farmers are unable to purchase water for cultivation from these water sellers, often without valid permits. Our research shows that since agriculture in this region provides very low profits, men migrate to distant towns and cities of West Bengal and other states in search of work. This situation has resulted in the feminization of agriculture and further impoverishment of marginalized communities (those with very low purchasing power parity, including low-caste Dalit Hindus, Adivasis [Indigenous peoples] and Muslim minorities, most originating from Bangladesh). The marginalized communities of this region are not only vulnerable but also have low resilience capacities. The character of land, water and labour along with other associated agriculture inputs have thus created a vicious cycle in which women, children and elderly people of the region are becoming victims of environmental stress, which is further aggravated during cyclonic events.

Access to private/public domestic water needs and clean drinking water varies and is poor, even though the Public Health and Engineering Department (PHED) of the Government of West Bengal (GoWB) stated that the 'Har Ghar Jal', Jal Jeevan Mission ('providing water to every household'), an initiative of the Indian government, is making good progress (Ministry of Jal Shakti, 2019; PHED, 2020). Water access is far from satisfactory; in most locations about 2000 inhabitants have to use only one community water tap connection. This has created a water emergency because there is inadequate water for bathing, cleaning, washing, rearing of domesticated animals, kitchen gardens, public water in schools and rural offices, local eateries, hotels, resorts and local manufacturing units.

According to the norms of water service provision by the Government of India, at least one handpump or piped water standpost is required for every 250 persons to meet the safe drinking water needs of the ISR. In addition, the Jal Jeevan Mission (2023) aims to provide safe and adequate drinking in rural India through individual household tap connections by 2024. However, in the ISR, surface water or pond water is used for washing, bathing and cleaning household utilities. Some people also use pond water for fish breeding aquaculture farms, irrigation for small plots, horticulture and other commercial purposes. Even though water supply varies with the seasons, contamination means that this water is not safe for domestic purposes. In fact, this surface water has become a breeding ground for mosquitoes and waterborne diseases. Unless proper pipeline systems are introduced in all villages, the use of deep tube wells remains the only source of drinking and cooking water in the region.

Refocusing attention: grassroots perspectives

Media reports affect perceptions, opinions and policy decisions taken by different government agencies and nongovernmental organizations (NGOs), who formulate various schemes and projects for the ISR. We analysed daily local newspaper articles published mostly from Kolkata and its surroundings in English and vernacular Bengali between 2012 and 2021. Many of these articles in *Sudhu Sundarban Charcha*, a quarterly publication dedicated to the study of the Sundarban, emphasize regional problems that focus on the environment, the socioeconomic situation, crimes, festivals, hazards, and disaster management (Lahiri, 2022). The environmental problems and natural hazards – especially cyclonic storms and the problems they bring, like sea water ingression due to embankment breaching, floods, and increasing soil salinity – are mostly highlighted in this magazine (Chakraborty, 2020).

Since 2009, climate change has been identified as the key factor for adverse consequences. Given the climatic hazards that occur in May and June, these are the months that receive the most interest, followed by October, August and December. In the period after the Aila Super Cyclone in 2009, audio-visual platforms and social media focused on 'natural' hazards, pollution levels and wildlife conflicts. Therefore, a critical understanding of media coverage in the Sundarban is crucial to understanding water access, equity and empowerment. The reports on access to improved quality of life and the development of sustainable water infrastructures are greatly lagging. Access to freshwater in the ISR is rarely the focus of print media reports. To supplement these understandings, Table 4.1 juxtaposes popular understandings from media reports with our findings from household surveys that convey the realities on the ground.

Table 4.1: Media representations and ground realities of water access in the Sundarban region

Media representation	Grassroots experience
As the Sundarban is in an estuarine region and criss-crossed by numerous creeks and channels, water is in abundance.	Creeks/river water mixed with saline sea waters at high tide become unusable for domestic and agricultural sector. This is exacerbated by storm surges and supercharged cyclones due to climate change-induced global warming.
Annual average rainfall of more than 1,700 millimetres sufficiently meets the water demand in the region.	Rainfall is seasonally skewed, and more than 80 per cent of annual rainfall occurs June–September; other seasons are mostly dry.
The presence of a large number of ponds and inland water channels suggests that modern irrigation is not necessary.	High evapotranspiration rates and low groundwater supply in the dry season occurs because the water table in most ponds and channels falls below the ecological requirement. Hence, judicious use through modern irrigation is a necessary supplement.
Groundwater is replenishable and is an unlimited resource.	Utilizable sweet water aquifers are too deep, and yields are low in April–May (summer) due to excessive pumping.

Source: Media reports and household field survey conducted by the authors

Stories from interviewees deepened understanding of water access and equity. In Jharkhali CDB, Siuli Majhi (aged 45), a female social worker responsible for women's and children's health in the Nehrupally village, voiced her view. She emphasized that although clean freshwater and hygienic sanitation is a prerequisite for good health, this requirement is not often met within any of the villages of the ISR. Another serious issue is the search for food, shelter and safety following disaster-induced displacements caused by devastating cyclonic events (Haldar et al, 2021a). For example, after Cyclone Aila, people from the Samsernagar village in Hingalganj CDB received food and drinking water supplies from the neighbouring villages of Koikhali in Bangladesh, when Indian government aid failed to reach the area in time due to its remoteness and inaccessibility (Biswas, 2022). These transnational solidarities are crucial to regenerative futures in the Sundarban, yet they remain undervalued in planning for safe and secure water futures.

Gender issues and climate change have always been a significant concern in the region (Haldar and Satpati, 2022). Tanuja Mollah (aged 32), a female social worker in Sonakhali village in Basanti CDB, commented on the health issues faced by women due to a shortage of qualified doctors. She also hinted at domestic violence, girl trafficking and other crimes as occurring in the region. Consequently, women with high inherent water resilience capacities had little opportunities to adapt. Pranati Jana (aged 48), a small motel owner from the Uttardanga village in the Gosaba CDB who runs/owns a successful business and is the sole earner in her family, shared her experiences. She spoke of frequent and recurring arguments at the public water pipeline standpost where about 50 households gather water during the stipulated flow time of three hours twice daily. Pranati estimated that each household needed at least three buckets (16 litres) of water, and women take around five to ten minutes to fill their containers. The time taken increases during the summer when the flow of water reduces; women stand in long queues in the hot sun for up to an hour or more. A total of 38 per cent of survey participants in both the Gosaba and Sagar CDBs complained of low flow rates, and 39 per cent said the taste was deteriorating due to salinity and other mineral contaminants (Bhattacharyya et al., 2019). Another 20 per cent said that even the colour of the water had turned reddish due to the high iron content (Field Notes, 2022). Other women in the same locality also complained of the salinity increase in pond water, particularly after cyclones, reporting the need to use groundwater for *Boro* rice cultivation, the winter crop when precipitation is negligible. Cultivation of *Aman* rice, the monsoon crop when excess precipitation occurs, is not possible in many places due to waterlogging.

From frequent field visits to this region, we observed that women from underprivileged households living in the remote islands of the ISR have to travel on an average 15 minutes on foot with heavy buckets to the community tube wells twice daily. They spend up to several hours a day fetching just enough water for their household consumption. The burden of water collection is the experience of women, which has repercussions on childcare and health (O'Brien, 2023). In addition, these women often do not have the option to build their own bathrooms for bathing and clothes washing, instead frequenting the local ponds and water bodies for these practices, which are often contaminated by organic pollutants and pathogens from chemical runoff. Adolescent girls need more water during menstruation, but this water is often unclean and can cause reproductive diseases. To protect their modesty, there are different bathing times for men and women at ponds. However, women respondents all wished for private toilets and bathing facilities in their households, which is still uncommon among the middle- or lower-class households in the villages we surveyed. Therefore, gender and socioeconomic class status continue to affect hygiene and access to proper sanitation.

Participants spoke of stress on water availability during events and religious festivals such as Makar Sankranti (Winter Solstice) that attracts more than a million people to Sagar Island; the largest island in the ISR. In January, visitors travel via Diamond Harbour and Kakdwip to Gangasagar temple at the southern edge of Sagar Island to take a holy dip, but en route they usually stay for a few days at various locations in the region, putting tremendous pressure on water and sanitation systems. During summer (May) and winter (December–January) foreign and urban tourists visit destinations in the Sundarban hoping to catch a glimpse of a 'Royal Bengal Tiger' in the wild. The popular spots, including Hingalganj, Gosaba, Pakhiralaya, Patharpratima and Bakkhali Beach, contain many tourist resorts, increasing demand for packaged drinking water, and often local brands have no quality control (Field Notes, 2022). A number of traders therefore operate drinking water businesses in the ISR, and even supply adjoining suburban areas of the Kolkata Metropolitan Area (KMA), such as Baruipur, Sonarpur, Diamond Harbour and Lakshmikantapur. Tourists often hire boats to travel along the rivers and creeks criss-crossing the interior of the region, and sufficient potable water is delivered to meet their needs during excursions. In many instances, used plastic bottles are thrown into the rivers, causing plastic pollution and environmental degradation. Moreover, this potable water industry has been found to create a conflict of interest between the local people and the suppliers of products in plastic packaging, many of whom do not stay permanently here. The profits from packaged water go into the pockets of the water traders and manufacturers, who have almost no concern for water equity in the region. In this context, women are found to be the worst sufferers and victims of water injustice (Field Notes, 2022).

Water management and empowerment

Whereas some of the more affluent families in the Sundarban have their own deep tube wells ensuring regular access to water, the relatively poorer sections of society have limited access to both standard-quality and a sufficient quantity of potable water. With more access to elementary and secondary education during the last decade or so (mainly due to the 'Kanyashree' Project of the West Bengal government, as observed from field responses), there is an increased awareness of personal health and hygiene, particularly among the younger generation. The traditional habit of using local ponds for cooking, washing or bathing is being discontinued, which has put more pressure on the local deep aquifers for domestic water. Some households have also tried to improve their access to drinking water by installing water filters. But again, the cost of maintaining these electronic devices, under high turbidity conditions, is high and economically unsustainable (Field Notes, 2022).

Therefore, it is clear that this region faces multifaceted water-related crises, including water shortages, unequal distribution of water and water deprivation, as well as a thriving water industry that is causing conflict among local users, water traders and local governments. Given this situation, it is necessary to explore the best possible management practices, keeping in mind the natural hydrological condition of the region and limited scope for successful human intervention. The ground realities like population growth, expansion of the urban agglomeration of the KMA towards the Sundarban, conversion of land use, poor water management, increasing tourist flow and the commercialization of farming, especially shrimp culture, must be duly acknowledged when preparing policy development plans. A holistic, well-planned and long-term strategy is needed for the region in order to ensure a future of adequate freshwater availability.

There are some promising state-led initiatives that are worth mentioning. For example, the West Bengal Government is implementing a programme to raise the height of tube wells to avoid submergence (Ministry of Water Resources, 2015). Some pilot projects have successfully supported artificial recharge of aquifers with rainwater to reduce the salinity. The Sundarban Development Board (SDB) is taking effective steps to develop water-related infrastructure in the region (Department of Sundarban Affairs, 2023). Approximately 10 million INR respectively had been earmarked under the twelfth and thirteenth National Five-Year Plans for surface water treatment plants in Saline Coastal Zones. Some solar pumps have been installed and a desalination plant with Reverse Osmosis (RO) purification techniques has been commissioned to provide saline-free water, which proved unsuccessful due to the high costs involved (Ministry of Water Resources, 2015).

We agree that it is difficult to arrest climate change-induced super cyclone hazards and related adversities at the microregional level of the Sundarban. Yet, preparedness for local-level adaptive measures focused on water equity and water justice can work. Our research suggests that strengthening local institutions – such as the *gram*/village panchayats, development authorities, as well as educational and research organizations – to form a synergistic framework for capacity building at the community level and among the vulnerable people of region is absolutely necessary. The development of indigenous adaptive technologies and changes in favour of environmentally friendly behavioural patterns will also be crucial. Providing equitable access to fresh water also requires assessing the usable water budget of supply and optimum demand at the community level, and based on participatory information, taking into consideration stakeholders' opinions that emphasize the experiences of women.

Techniques of rainwater harvesting and catchment management are recommended where and when there is adequate rainfall. Our research suggests that extensive rainwater harvesting must be adopted with the help of proper infrastructure, since the region receives more than 170 centimetres of average annual rainfall. Our field surveys reveal that the cost of buying water in the summer is about INR 100 for 20 litres/day for a 60-day period, that estimates to INR 120,000 per household, especially in April and May. Another INR 500 per month is incurred for sweet water used for other purposes such as agriculture and domestic use. Hence, the cost of buying water over eight years is roughly estimated at INR 1 million. The cost of installing shallow pumps to irrigate 324.3 hectares is about INR 70,000. If, on the contrary, rainwater harvesting (RWH) is practised, then the cost of constructing a RWH pond (150 square metres) which can sustain more than ten people for a minimum of eight years would be no more than INR 25,000. Yet this requires proper planning and implementation of water policies, as well as the appropriate development and maintenance of water infrastructure for the sustainable development of the region.

Regular monitoring of land use and land cover by using historical spatial datasets of different time windows is helpful in terms of estimating water demand in the region. The principle of social justice or equitable access to, and distribution of, locally available resources should be followed to enable the participation of marginalized groups (especially women) in decision-making processes for the management and sustainable development of local resources. In this regard, it is worth recalling that Article 21(A) of the Indian Constitution recognizes the provision of potable water to every citizen of the country as a fundamental right. However, our research has highlighted that while there is an awareness of water issues and problems, there is still a relatively low level of political, economic and technological understanding needed for developing sustainable solutions among communities on the frontline of water stress. Many marginalized people, especially women, struggle with agencies that control and provide water resources. Moreover, when water supply systems are installed in a top-down approach with no meaningful consultation and education, the infrastructure is very often inadequately looked after, and in the event of damage, restoration of the water supply takes many weeks, causing severe hardship to women in the region.

At the grassroots level, *gram* (village) panchayats, formal institutions formed under the West Bengal Panchayati Raj Act, 1973, and subsequent national amendments (see the 73rd Constitution of India (Amendment) Act, 1992) suggest that local governments have the power to control and manage local resources. Our fieldwork suggests that the *gram panchayats* can play a significant role in providing clean water to the local communities for various purposes. Under the Jal Dharo, Jal Bharo (Catch water, Store Water) scheme (Department of Information and Cultural Affairs, 2022), many old ponds have been renovated, resuscitated and new ones have been dug. Yet the water quality management of the surface systems has not been as successful, due to a lack of adequate informed participation from local communities. Moreover, corruption within government machinery and malpractices within various sanctioned schemes by influential local leaders jeopardize the very purpose of the projects (Group Interview, Field Survey, 2022). Under the Mahatma Gandhi National Rural Employment Guarantee Act (MGNREGA), 2005, within the Ministry of Rural Development, various schemes have been introduced that provide the rural population with a minimum of 100 days' work (Desai et al, 2015). Within the ambit of this programme, there are various water harvesting projects that can store more surface water during the monsoon season, including the excavation of tanks, and the renewal of silted-up ponds and decaying channels. This water can then be made available for *Boro* rice cultivation to reduce the pressure on groundwater. Maintaining supply systems and managing water requires the regular monitoring of water needs and water development projects, and, crucially, the help of local people. Water education for capacity building and social empowerment is possible through water management by local communities.

Conclusion

The Indian Sundarban Region is a highly sensitive and fragile tidal ecosystem where land, water, flora, fauna and human beings with their local cultural attributes have formed a unique ecoheritage site. But over time, the region has been facing the challenges of rapid transformation. Local people, in particular the poorer and marginalized sections of society, are becoming more impoverished. Water injustice is at the centre of these pressing problems. Non-equitable distribution of freshwater has created water crises for millions of people in the Sundarban, a situation that can be further aggravated by water traders' control of water supply systems. It is vital that an equitable participatory management system that focuses on water is developed in the region, with the local government (rural *panchayat*) playing a significant role.

The local people of the Sundarban are more concerned about meeting their immediate requirements that sustain their livelihoods; the idea of 'thinking globally' is not quite relevant to their day-to-day affairs. However, they are affected by the various physical and socioeconomic interconnections that have long-term impacts on their lives. Scientists and academics should partner with local communities in creating collaborative understandings of various environmental issues, such as water access and equity. Through effective communications and use of collaborative knowledge, tools and techniques, better freshwater management is possible. Academics have a responsibility to share innovative ideas with local governments and voluntary organizations to produce more effective solutions for local problems.

The ongoing water crisis and the hegemonic control over the distribution of freshwater in the ISR have become more complicated by climate change-induced hazards affecting the region and its marginalized people who are not directly responsible for global environmental changes. Equitable reallocation of water access is an expression of environmental justice that ensures the right to life, as enshrined in the Article 21 of the Indian Constitution (O'Brien, 2023). Further, Article 14(2)(h) of the Convention on the Elimination of All Forms of Discrimination against Women (CEDAW) reads: 'States parties shall take all appropriate measures to eliminate discrimination against women in rural areas in order to ensure, on a basis of equality of men and women, that they participate in and benefit from rural development.' But the burden of water collection often falls on women who travel long distances and stand in long queues to fetch water. The emergent issues for which policies need to be implemented in a bottom-up manner includes increasing participation in the education sector, decent work opportunities, childcare, preventing exposure to verbal, sexual and physical violence and discrimination, and equitable communities all of which are directly or indirectly interconnected with water security in the ISR.

Note
[1] The authors are immensely grateful to the people of the Indian Sundarban Region who have consented to narrate their water challenges. A special mention to Dr M. Lobo, School of Humanities and Social Science, Deakin University and the editors, reviewers and all associated in refining the chapter and giving their valuable input.

References

Bhadra, T., Das, S., Hazra, S. and Barman, B.C. (2018) 'Assessing the demand, availability and accessibility of potable water in Indian Sundarban biosphere reserve area', *International Journal of Recent Scientific Research*, 9: 25437–25443.

Bhadra, T., Hazra, S., Roy, S.S. and Chandra Barman, B. (2020) 'Assessing the groundwater quality of the coastal aquifers of a vulnerable delta: a case study of the Sundarban Biosphere Reserve, India', *Groundwater for Sustainable Development*,: 100438, DOI: 10.1016/j.gsd.2020.100438.

Bhattacharyya, A., Haldar, A., Bhattacharyya, M., & Ghosh, A. (2019). Anthropogenic influence shapes the distribution of antibiotic resistant bacteria (ARB) in the sediment of Sundarban estuary in India. Science of the total environment, 647, 1626-1639.

Biswas, C. (2022) 'Potable water crisis and the Sundarbans', *Borderless Journal*, 14 January, Available from: https://borderlessjournal.com/2022/01/14/potable-water-crisis-the-sunderbans/ [Accessed 12 March 2024].

Census (2011) 'Primary Census Abstracts', Registrar General of India, Ministry of Home Affairs, Government of India, Available from: https://censusindia.gov.in/census.website/data/population-finder

Chakraborty, S. (2020). 'Rising drinking water salinity affecting health, livelihoods in Sundarbans region', *newsclick.in*, Available from: https://www.newsclick.in/Rising-Drinking-Water-Salinity-Affecting-Health-Livelihoods-Sunderbans-Region [Accessed 15 January 2023].

Das, K., Mukherjee, A., Malakar, P., Das P. and Dey U. (2021) 'Impact of global-scale hydroclimatic patterns on surface water-groundwater interactions in the climatically vulnerable Ganges river delta of the Sundarbans', *Science of the Total Environment*, 798: 149198.

Department of Information and Cultural Affairs, (2022) 'Schemes – Jal Dharo', Government of West Bengal, Available from: https://wb.gov.in/government-schemes-details-joldharo.aspx [Accessed 3 April 2023].

Department of Sundarban Affairs (2023) 'Initiatives taken by Sundarban Affairs Department in 2012–2013 and 2013–2014', Government of West Bengal, Available from: https://www.sundarbanaffairswb.in/home/page/initiatives#3 [Accessed 3 April 2023].

Desai, S., Vashishtha, P. and Joshi, O. (2015) 'Mahatma Gandhi national rural employment guarantee act: a catalyst for rural transformation', *eSocialSciences*, Working Paper ID: 7259.

Do Amaral, P.S., Zanatta, F., Meireles, G.B., Mendes, J.P., de Assis Cosso, S.P. and Mariosa, D.F. (2023) 'Fundamentals and ethical consequences of applying the principles of universalization, sustainability and water safety contained in the Regulatory Framework for Sanitation in Brazil', *Gestão & Regionalidade*, 39, DOI: 10.13037/gr.vol39.e20237565.

Haldar, A. (2015) 'Trend analysis of cyclones, over Kolkata and South 24 Parganas Districts, West Bengal', in A. Haldar and L. Satpati (eds) *Climate and Society – A Contemporary Perspective*, Kolkata: University of Calcutta, pp 49–60.

Haldar, A. and Debnath, A. (2014) 'Assessment of climate induced soil salinity conditions of Gosaba Island, West Bengal and its influence on local livelihood', in *Climate Change and Biodiversity* (Proceedings of IGU Rohtak Conference, Vol. 1), Tokyo: Springer, pp 27–44.

Haldar, A. and Satpati, L. (2022) 'Climate change hazard perception and gender empowerment for resilience in the regions of Tropical Asia', in R. Singh, K. Ram, C. Yadav and A.R. Siddiqui (eds) *Climate Change, Disaster and Adaptations: Contextualising Human Responses to Ecological Change*, Cham: Springer International Publishing, pp 73–82.

Haldar, A., Das, S., Chatterjee, R. and Satpati, L. (2021a) 'Environmental vulnerability and displacement due to land erosion: selected case studies in West Bengal, India' in A. Haldar, A. Alam and L. Satpati, L (eds) *Habitat, Ecology and Ekistics: Case Studies of Human-Environment Interactions in India*, Cham: Springer, pp 207–224.

Halder, S., Kumar, P., Das, K., Dasgupta, R. and Mukherjee, A. (2021b) 'Socio-hydrological approach to explore groundwater–human wellbeing nexus: case study from Sundarbans, India', *Water*, 13(12): 1635, DOI: 10.3390/w13121635.

Hazra, S., Bhadra, T. and Ray, S.P.S. (2019) 'Sustainable Water Resource Management in the Sundarban Biosphere Reserve, India', in S. Ray (ed) *Ground Water Development: Issues and Sustainable Solutions*, Singapore: Springer, pp 147–157.

Lahiri, J. (2022) *Sudhu Sundarban Charcha*, volumes 11 and 12 (April–July), West Bengal: Hooghly, pp 1–104.

Ministry of Environment, Forest and Climate Change (Government of India) (2019) 'Ministry Of Environment, Forest and Climate Change, Notification' *The Gazette of India*, 18 January, Available from: http://environmentclearance.nic.in/writereaddata/SCZMADocument/CRZ_Notification2019.pdf [Accessed 15 January 2023].

Ministry of Jal Shakti (2019) *Jal Jeevan Mission: Har Ghar Jal. Reforms in Rural Drinking Water Supply*, Department of Drinking Water and Sanitation, Government of India, Available from: https://jaljeevanmission.gov.in/sites/default/files/manual_document/JJM-Reform-Document-English.pdf [Accessed 3 April 2023].

Ministry of Water Resources (2015) 'Water supply position in Sunderban', Press Information Bureau, Government of India, Available from: https://pib.gov.in/newsite/PrintRelease.aspx?relid=121357 [Accessed 3 April 2023].

Mitra, A., Gangopadhyay, A., Dube, A., Schmidt A.C.K. and Banerjee, K. (2018) 'Observed changes in water mass properties in the Indian Sundarbans (northwestern Bay of Bengal) during 1980–2007', *Current Science*, 97(10): 1445–1452.

Ministry of Drinking Water and Sanitation (2018) 'Swajal: a community-led approach to rural piped drinking water supply. Guidelines', Ministry of Drinking Water and Sanitation, Government of India.

O'Brien, D. (2023) 'Opinion: news channels, chasing TRPs, ignore the necessary', *NDTV*, 3 January, Available from: https://www.ndtv.com/opinion/news-channels-chasing-trps-ignore-the-necessary-3659113 [Accessed 12 March 2024].

Public Health Engineering Department (PHED) (2020) 'Vision 2020', Public Health Engineering Department, Government of West Bengal. Available from: https://wbphed.gov.in/en/pages/vision-2020 [Accessed 3 April 2023].

Rockström, J. and Gordon, L. (2001) 'Assessment of green water flows to sustain major biomes of the world: implications for future ecohydrological landscape management', *Physics and Chemistry of the Earth, Part B: Hydrology, Oceans and Atmosphere*, 26(11–12): 843–851.

Rudra, K. (2019) 'Interrelationship between surface and ground water: the case of West Bengal' in S. Ray (ed) *Ground Water Development: Issues and Sustainable Solutions*, Singapore: Springer, pp 175–181.

Seitzinger, S.P. et al (2012) 'Planetary stewardship in an urbanizing world: beyond city limits', *Ambio*, 41(8): 787–794.

UNICEF (2021) *The Climate Crisis Is a Child Rights Crisis: Introducing the Children's Climate Risk Index*, New York: United Nations Children's Fund.

United Nations (2022) 'Goal 6: Ensure availability and sustainable management of water and sanitation for all', United Nations Department for Economic and Social Affairs, Sustainable Development, Available from: https://sdgs.un.org/goals/goal6 [Accessed 2 March 2023].

Vajpeyee, G. (2023) 'Low oxygen levels in Pune rivers put aquatic life at risk: report', *Hindustan Times*, 3 April, Available from: https://www.hindustantimes.com/cities/pune-news/pune-rivers-dissolved-oxygen-below-standard-55-polluted-river-stretches-in-maharashtra-cpcb-report-101680368681954.html [Accessed 2 April 2023].

World Bank (2014) *Building Resilience for Sustainable Development of the Sundarbans through Estuary Management, Poverty Reduction, and Biodiversity Conservation*, World Bank Report No. 88061-IN, 273.

5

Climate Change and Oceanic Responsibilities: Listening and Dancing with Saltwater Country, Australia

Lowell Hunter and Michele Lobo

Decolonial crosscurrents

We express our gratitude and respect to past, present and emerging knowledge holders as well as ancestral spirits of the Southern and Indian Ocean. Their ways of being and becoming with Sea Country or Saltwater Country that is woven with 'sacred energies', songs, feelings and stories enable us to tread lightly, voyage carefully and belong amid anthropogenic climate change (Buku-Larrngay Mulka Centre, 1999, p 54; Lobo et al, 2022). As we walk along the shore, wade, venture out in boats or on surfboards in Gunditjmara and Wadawurrung Saltwater Country in southern Victoria, Australia, we are drawn into the kinetic energy of the planet that mobilizes wind systems, cyclones, anticyclones, ocean currents, waves and tides. This energy materializes in coastal places when sand drifts, dunes slip, rocky cliffs weather, erode, recede; and birds, fish, reptiles as well as marine mammals (including whales) feed, breed and migrate. We feel these sacred and kinetic energies as we come together to create decolonial crosscurrents in Saltwater Country strangulated by global warming, acidification, sea level rise, pollution and offshore energy extraction.

Sharing stories as a Nyul Nyul saltwater man and sand artist Lowell Hunter from the small coastal town of Broome in Western Australia and a geographer from the deltaic city of Kolkata, India, our Indigenous-southern decolonial praxis is a creative alliance that aims to develop and stretch embodied oceanic literacies of responsibility from and beyond settler colonial Australia. These literacies emerge from sensing, listening, moving and dancing with the sacred and kinetic energies of the Southern Ocean. We build on Kanaka Maoli (Native Hawaiian) scholar and surfer Karin Ingersoll's (2016) seascape epistemology in the Pacific Ocean that creates waves of knowing through embodied encounters. Our multisensory, visceral and immersive encounters with the cool Southern Ocean, a home away from home on the Indian Ocean edge, drives our politics that decentres disembodied critical debates on racial justice, climate justice, multispecies justice and planetary justice led by 'Western scholars and elites' (Okafor-Yarwood, 2022, p 2).

Writing from the Global South, Okafor-Yarwood (2022) forcefully argues that marine vulnerabilities and inequities cannot be addressed from a Western or elite level of consciousness that has contributed to the creation of racial injustice and oceanic injustice. For us, a blackfella and a brown immigrant woman, a decolonial praxis of solidarity affirms oceanic responsibilities and unsettles the language of climate crises, *aqua nullius* and 'white eco-apocalyptic futures' (Gergan et al, 2018, p 15). Yet, Tuck and Yang (2012) remind us of the risks of domesticating decolonization in settler colonial nations such as Australia, where white innocence and white futures are easily recentred.

First, this chapter mobilises decolonial crosscurrents of dreaming, reinvigoration, and healing with Saltwater Country amid documented scientific literacies of anthropogenic climate change and the sixth mass extinction. These decolonial political crosscurrents cannot flow, swirl or strengthen, however, if Indigenous and Southern thinking is trapped in a negative critique of climate colonialism and climate justice. Cultural geographers in the geohumanities argue for an 'affirmative critique' (Squire et al, 2022, p 519) that galvanizes people to come together with enthusiasm to engage with ethical complexities of the Climate Emergency in creative ways. In response to anthropogenic climate change, strangulated ocean spaces and dystopic futures, we have a responsibility to do more than critique the One-World World (OWW) that privileges the 'capitalist, patriarchal, colonial, globalized world' (Escobar, 2020, p 9) and the brutal racial logics of 'human' mastery and white supremacy. Such work is affective labour, particularly in 'climate hotspots' of the Global South, including the low-lying islands of Tide Country (*bhatir desh*) in the UNESCO heritage-listed Sundarbans mangrove forest in deltaic Bengal, south of Michele's 'home' in Kolkata. This is a place where top-down state-corporate collusions focus on climate solutionism, but exclude the poor who are racialized, displaced and dispossessed (Lobo et al, 2022; Srivastava et al, 2022; Chapter 4 in this volume). Indigenous, Southern and Black-led abolitionist thinking on 'constellations' of co-resistance (Gilmore, 2022, p 3; Maynard and Simpson, 2022), 'beautiful experiments' (Hartman, 2019, p 46), 'rehearsals for living'

(Gilmore, 2022, p 3) and 'sand talk' (Yunkaporta, 2019, p 17) enable us to hear the chorus of responsibility for an earth otherwise that rings out from human and more-than-human ocean spaces.

Second, this chapter visualizes and vocalizes the chorus of oceanic responsibility from Gunditjmara and Wadawurrung Saltwater Country along the Southern Ocean. For First Nations peoples, sea, land and sky is woven with 'sacred energies' (Buku-Larrngay Mulka Centre, 1999, p 54), laws, creation stories and songlines of ancestral beings, broken by climate change and continuing threats of extractive capitalism and fossil-fuel energy futures (Bradley, 2010, Bundle et al, 2022). Songlines or songspirals are like maps that weave relational becoming with Country (see Chapter 2) that is alive, sentient and vibrant (Gay'wu group of women, 2019). Lowell's experimental sand artworks at the beach, our collaborative video artwork (Hunter and Lobo, 2022), and his participation in an Indigenous ceremony at a climate change activist event held on Earth Day, demonstrate how relational becoming with Saltwater Country is performed, strengthened and communicated.

Indigenous-led activism by the Southern Ocean Protection Embassy Collective (SOPEC) focuses on protecting whale songlines in Gunditjmara and Wadawurrung Sea Country from environmental damage (Bundle et al, 2022). National and international tourists who travel along the winding, scenic, heritage-listed Great Ocean Road may not always attune to these songlines that weave relational becoming through land and sea. They are lured to this beachscape with rugged cliffs, golden sandy beaches (such as Torquay, Bells, Lorne and Apollo Bay), ethereal sunsets and limestone stacks (Twelve Apostles, Port Campbell National Park). In autumn, they celebrate the gigantic ocean swells of Bells Beach during the Rip Curl Pro, an international professional surfing competition. Further west, near Warrnambool, wildlife enthusiasts and whale watchers congregate at the viewing platform at Logan's Beach. They watch and wait patiently with curiosity hoping to see southern right whales migrate north to breed in warmer waters off the coast of Victoria, from May to September. Scientific research, however, shows that whale songs, communication and navigating capacities between feeding and breeding grounds are affected by anthropogenic climate change as well as marine 'seismic survey sounding' for oil and gas exploration (Warren et al, 2020, p 9). It is therefore not surprising that there were more than 2,000 submissions by youth, community groups, and Indigenous peoples to the State and Commonwealth governments opposing Viva Energy's proposed floating offshore liquefied natural gas terminal in the industrial area of Corio Bay, Geelong, in southern Victoria (Clayton, 2022).

Third, this chapter draws attention to everyday political activism that attunes to the more-than-human 'life force' of Saltwater Country (Buku-Larrngay Mulka Centre, 1999, p 105), and strays from rehearsing the threats to ocean spaces highlighted in scientific, policy and media narratives. In Australia, this life force was first embodied in sacred designs (*miny'tji*) painted on tree barks in the 1990s by the Madarrpa and Gumtaj clans in northeast Arnhem Land (northern Australia). Rather than respond with grief and anxiety to the desecration of the ancestral home of the crocodile Baru – a 'primal force' in Blue Mud Bay – Indigenous clans affirmed this life force in beautiful paintings (Buku-Larrngay Mulka Centre, 1999, p 9). Wulamba or the roaring ocean with ancestral beings such as crocodiles and turtles was etched by the smell, sound, taste and the touch of the sea. This sacred art was used by Yolgnu to argue for Sea Rights (in Balanda/white settler language), which were finally recognized in 2008. Like sacred designs on bark, Lowell's sand art visualizes and vocalizes the spiritual and physical touch of Bunjil the creator spirit, ancestral beings such as the whale as well as the kinetic energy of currents, waves, tides and winds along the winding heritage listed Great Ocean Road in Southern Victoria. Expressions of this multisensory and spiritual touch are beautiful and generous gifts amid apocalyptic narratives of climate change and extinction that gather force, become contagious and, as Baldwin (2016) suggests, risks recentring white affects of eco-anxiety, grief and despair.

Over the last three years, our face-to-face meetings, telephone conversations, Instagram messages as well as the sharing of drone imagery, stories and yarns inspired our 'beautiful experiments' (Hartman, 2019, p 46) of multisensory storying with Gunditjmara and Wadawurrung Sea Country along the Great Ocean Road. These experiments provide a new perspective to scientific and policy literacies of anthropogenic climate change, heritage loss and biodiversity conservation, advanced by diverse stakeholders including the Victorian Department of Energy, Environment and Climate Action (DEECA), Great Ocean Road Coast and Parks Authority (GORCAPA) and local councils (Surf Coast, Colac Otway, Corangamite and Warrnambool). Along with Aboriginal custodians and stewards, we want to respond to the call: 'Paleert Tjaara Dja – Let's make Country good together again' (Wadawurrung Traditional Owners Aboriginal Corporation, 2020).

Oceanic literacies

Sand art: dancing responsibility

Lowell: I go down to the beach and create sand artworks. Growing up in Broome in the Kimberley regions of Western Australia I remember, being a young boy, being with my family and understanding that we had this relationship with the sea and that everything revolved around that – so life revolved around living by the Indian Ocean. Moving to Victoria in southern

Australia I still had this urgency and this need to be by the ocean. I was fortunate enough to grow up in Warrnambool along the Southern Ocean and nourish that strong association and connection to the sea. It wasn't just a personal connection, as a saltwater man I also had an environmental connection that was cultural and spiritual.

I fast forward to where I am now. I have danced traditionally for about 25 years and that's been a big part of my journey, and it's actually influenced a lot of the work that I do today. So, the artworks that are created on the beach represent my lived experiences as an Aboriginal man. I talk and create stories that represent my family, my community, and different elements of our culture. Really, drawing on that relationship with the ocean that I've lived with and live with now, and wanting to share that with my children as well. So, generally my artwork involves working with the ocean and working with Country and understanding that I have this relationship with it. You can't go down there [the beach] every single day and do what I do. It's understanding the tides, understanding when the wind comes in, because I use drone technology to create my artworks; it's really dependent on what's happening in Country. When I think about our First Nations people's relationship to Country, my art very much revolves around that relationship. Being interconnected with the environment, and not positioning myself above it (Figure 5.1). It's really about living harmoniously with the land and with the sea in order to get a life of fulfilment and connection, spiritually and culturally.

Figure 5.1: Connection to Saltwater Country

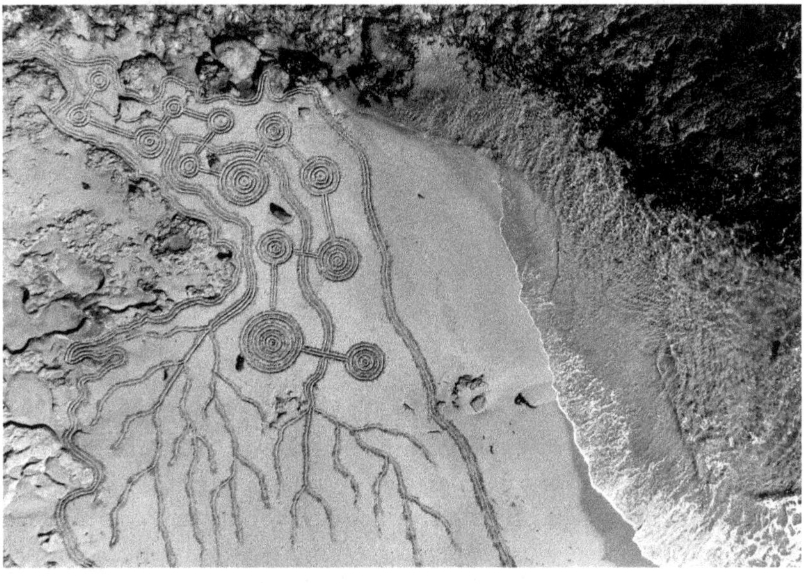

Source: Lowell Hunter, 5 April 2020

Using only my feet, I carve these stories of connection into the sand using the same foot movements I was taught through traditional dance movements my people have practiced for countless generations. I etch transitory sand artworks visible for a few hours before they get blown away by gusty southern winds or washed away by high tides. I remember posting this comment on Instagram after completing an artwork at Thirteenth Beach along the Great Ocean Road, 'Nature Wins – Lucky I had captured this artwork before this happened' [28 June 2020]. I try and make sensible the diurnal and seasonal rhythms of Gunditjmara and Wadawurrung Sea Country at Barwon Heads, Thirteenth Beach, Bancoora Beach, Torquay Beach, and Warrnambool Beach. These rhythms that bring together sacred and kinetic forces are partially captured by my drone photography through artworks titled Whale Dreaming, Reinvigorate and Heal Country. These artworks go viral through social media [37,400 followers in March 2024] and 'help me find my way' amid the violence of anthropogenic climate change. I also print these artworks, exhibit them in local community galleries and sell them online.

Whale Dreaming celebrating whale songlines

Lowell: Whale Dreaming is a story about me growing up down in southwest Victoria and learning traditionally the whale dance of the Gunditjmara People. So, we performed whale dances on heaps of occasions growing up, and then later in life when I started to do my sand art. When I went down to the ocean, I never really thought about what I was going to create that day. It just came to me – I just thought well whales actually travel through this Country and also travel up to where my people are originally from in the Kimberley. So, I created this artwork Whale Dreaming (Figure 5.2) that is inspired by my whale sightings along the Great Ocean Road. Whale Dreaming is special because it renews intimate personal relationships with my ancestral Nyul Nyul Sea Country in Beagle Bay, north of Broome in Western Australia.

Michele: Between June and September endangered humpback whales, southern right whales, blue whales, orcas (killer whales), sperm whales and pilot whales journey north from Antarctica to the warmer Southern and Indian Ocean. These cetaceans or aquatic mammals that connect Lowell to the Kimberley also connect me to the Bay of Bengal in the Indian Ocean south of Kolkata. I was recently [20 June–September 2021] deeply affected when the carcasses of two whales were washed ashore along the bay at Digha [Mandarmani] and Bakkhali beaches, which I know so well. This was followed by a mass whale stranding off the western coast of Tasmania, Australia in September 2022; 270 pilot whales [each weighing approximately

Figure 5.2: Whale Dreaming

Source: Lowell Hunter, 23 April 2020

2,300 kilograms] were stranded, most died (Brown, 2022). These strandings so close to home in Kolkata and Melbourne encourage dreaming about kinship with whales.

Lowell: A year after creating my Whale Dreaming artwork at Bancoora beach on Wadawurrung Country, I posted this comment on Instagram on 23 April 2021 along with a photograph/video of the artwork captured by a drone. It read: 'I remember going down to the beach that day not knowing what I was going to create, then I had something come to me – in my mind, in my body and in my spirit … so I created what was talking to me.' My dancing that produced the sand artwork Whale Dreaming inspired me to engage in political activism to protect *koontapool* (southern right whale) songlines. Along with old friends and Elders, I galvanized people who live in Gunditj Country to sign a Citizens' Protection Declaration prepared by Traditional Custodian family clans and allies of other cultures (Figure 5.3). I sent the declaration to Michele for her signature on Instagram, a social media platform that I use frequently as 'saltyonehere'.

The declaration, an Indigenous-led response to climate change, focuses on ocean conservation, protection, education, as well as ecotourism and refuses the expansion of fossil fuel projects. It calls for listening to the oldest storytellers and demands negotiation so that Sea Country in southwestern Victoria can be cared for and protected. The declaration reads:

> *Wantayngeenkopa leekanyoong ngootook*?? What the matter with you that you destroy future generations right to live?? Mine and

future generations have inherited a world of pollution, disease and mass extinction.

Vicky Couzens, Gunditjmara Keeray Whurrong Elder

On Earth Day, 5 June 2022, Yaraan Couzens Bundle, a Gunditjmara childhood friend and founder of The Southern Ocean Protection Embassy Collective [SOPEC] invited me to dance at a ceremony to honour our whale ancestors.

Figure 5.3: Citizens' Protection Declaration

Source: Lowell Hunter and Michele Lobo

Michele: SOPEC is a Gunditjmara First Nations People's Ocean Conservation and Climate Solutions not-for-profit organization, committed to protecting the whale [*koontapool*] birthing and songlines across the Great Ocean Road (Bundle et al, 2022). The Indigenous-led activist organization brings together Traditional Owners as well as local citizen groups and hosts a Facebook page where photographs, videos, announcements and comments are posted. I watched a video of the ceremony held at Harris Reserve close to the Logan Beach whale nursery in Gunditjmara Sea Country. A group of 20 women and children dressed in black, holding eucalyptus branches, danced moving in an anticlockwise circle around a large cloth banner inscribed with Aboriginal languages and bordered by red handprints. Images of the ocean, sun and sky on the banner symbolized the coming together of Earth, Sky, and Sea Country.

Lowell: With my body painted in red and white ochre clay, I danced with four men and two children around the cloth banner to Aboriginal singing and music we created by hitting together two painted wooden clapsticks. The dance was followed by burning eucalyptus leaves in a smoking ceremony. The sacred ceremony concluded with outstretched hands and crouching bodies of women, men and children welcoming and inhaling the reinvigorating smoke.

Lowell: Reinvigorate is a significant work for me, and the cultural stories that live within Reinvigorate (Figure 5.4) revolve around the dispossession

Figure 5.4: Reinvigorate

Source: Lowell Hunter, 2 August 2020

Figure 5.5: Saltwater Healing

Source: Lowell Hunter, 31 May 2021

and disconnection that we have had from culture, due to the invasion and colonization of Australia by white settlers; not being able to, I guess, connect with culture in the way that we would have traditionally because of the massacres, the loss of language, the loss of culture. So, the artwork, Reinvigorate is me holding my power as an Aboriginal man, understanding the connection that I have to my culture, drawing on that, and bringing that back into this world that I live in now. And, understanding that even though my Elders and my ancestors are not there with me right at this moment, I know spiritually they are there to guide me and protect me. That's really what the reinvigoration of culture is to me as well. It mobilizes saltwater healing.

Lowell: Saltwater Healing (Figure 5.5) is an artwork and exhibition held in 2020 that embodies dreams of reinvigorating but also healing Sea Country. I danced in circles for several hours to create the sand artwork along the shoreline.

Michele: The circles produced through Lowell's dancing feet are comparable to bubbles of 'life force' produced by whales as well as sacred clan designs [*miny'tji*] in the Yirrkala bark paintings of Sea Country in north-east Arnhem land (Buku-Larrngay Mulka Centre, 1999, p 105). A year later, Heal Country, Heal Climate was a statement developed and led by a national coalition of three hundred Aboriginal and Torres Strait Islander organizations in Australia to inform environmental priorities at the 26th United Nations

Climate Change Conference [COP26] held at Glasgow in 2021 (Indigenous Peoples' Organisation-Australia, 2021).

Video artwork: listening and dancing with Saltwater Country

Lowell and Michele: We co-produced a video artwork that advanced our thinking on reinvigorating and healing sea country. The artwork, titled 'Climate Change and Oceanic Responsibilities: Listening and Dancing with Saltwater Country' (Hunter and Lobo, 2022), emerged through sharing stories of our ancestral country along the Indian Ocean where knowledge of the sea supports precarious livelihoods affected by floods, sea level rise and devastating tropical cyclones. The video and oral paper that primarily focused on Saltwater Country along the Great Ocean Road was presented at the Association of American Geographers Conference in February 2022, the largest international geography conference, attracting more than 9,000 participants. Lowell's participation in the session, titled 'Crosscurrents of Oceanic Thinking', was supported by the award of a prestigious Enrichment Grant that recognized the sociocultural, political and environmental benefits of decolonizing the white space of the academy through Indigenous and Southern insights into multispecies and planetary justice.

The video [2 minutes, 19 seconds] opens with a drone view of the bare back of a lone Indigenous dancer treading lightly on the wet sandy edge of the Southern Ocean at sunrise. Moving toward the ocean he claps together two boomerangs to the sound of the beat of electronic music. This moving imagery of brown, gold and blue morphs into a close-up view of a beach artwork with the black, red and yellow Aboriginal flag at the centre; Lowell stands there smiling and holding a small drone. The drone view zooms out to include the wide sandy beach, the busy arterial road over the Southern Ocean near the regional city of Geelong and clumps of green trees. Then we walk with Lowell towards the water's edge with submerged rocks and dance with him on the sand as he moves and creates the artwork Whale Dreaming. As waves lap against the shore and break, the artwork is partially erased. This ceremony of listening and dancing with Sea Country becomes a collective experience when five men with bodies painted in white ochre clay move away from the wet sandy edge and sit with eyes closed, feeling, breathing, and inhaling the salty air from Saltwater Country. Reinvigorated, they stand still on the dry sandy beach, but soon their tiny soft footprints on the wet sand and long shadows are only visible in the distant drone view (Figure 5.6). The video ends with a ceremonial dance and the clapping of boomerangs on the wet sandy edge that seems to suggest emergence and bodily rhythms with the waves of the Southern Ocean. This presentation at a large international geography conference straddled the boundary between art, activism and the academy, and in creative ways unsettled the anthropocentric

Figure 5.6: Video artwork

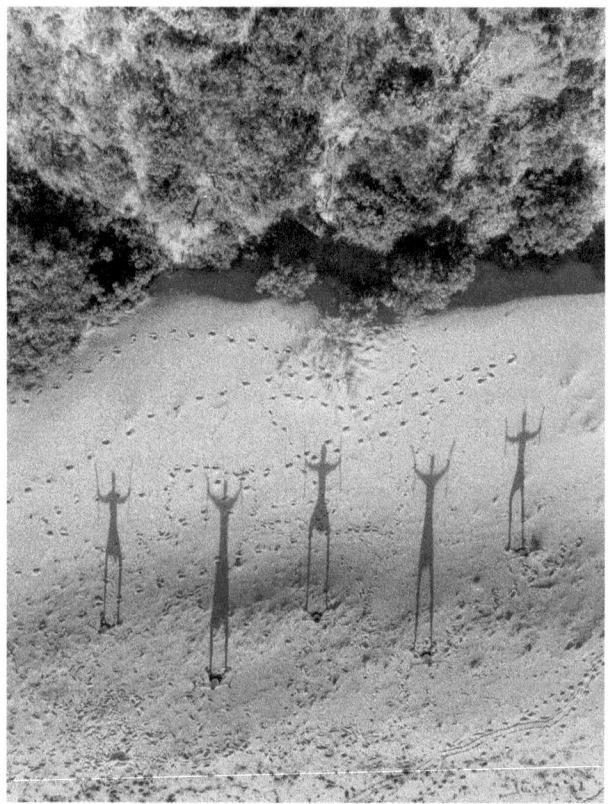

Source: Lowell Hunter, February 2022

view of the Australian beach as a white space of possession, mastery, freedom, adventure and a 'laid-back' lifestyle (Lobo, 2014; Moreton-Robinson, 2015).

Oceanic responsibilities: multisensory storying

Michele and Lowell: As an Australian woman of Indian heritage and a Nyul Nyul saltwater man, this chapter is a humble and respectful response to the invitation by Traditional Owners. They invite us to live by the lore of Bunjil, the creator spirit and make Saltwater Country healthy and good together through values that are cultural, ecological, economic and livelihood-based (Wadawurrung Traditional Owners Aboriginal Corporation, 2020). As we think of the torrential rain and recent floods in Lowell's home in Broome and the unseasonal cyclones such as Amphan that washed away lives in the Sundarbans and devastated deltaic Bengal in May 2020, with wind speeds of over 180 kilometres per hour, we know the pain of embodying an environmental ethics of listening and dancing with Saltwater Country.

Our oceanic literacies are embodied, visual, affective, kinaesthetic and spiritual, and have the potential to produce new kinetic energies of Black responsibility. Karin Ingersoll (2016), an Indigenous scholar, draws on Kanaka Maoli [Native Hawaiian] ontologies of the Pacific Ocean to argue for oceanic literacies or alternative relationships and interconnections between the sky, ocean, land, plants, animals, celestial bodies and the cosmos in the Anthropocene. Sand art or patterns that materialize dreaming, healing and reinvigoration take Kanaka Maoli oceanic literacies of more-than-human multisensory relationships, kinships and connections in new directions. Yunkaporta (2019, pp 17, 19), who belongs to the Apalech clan in Western Cape York, northern Australia, argues that 'Indigenous pattern thinking' or 'sand talk' is important in visualizing Indigenous Knowledge perspectives on more sustainable systems, and invites deep engagement rather than just polite Acknowledgement of Country. This chapter opens to these new patterns of creative co-resistance enabled by digital sensing technologies that listen and dance with Saltwater Country in the Anthropocene.

Oceanic turns that share Aboriginal and Southern rituals and laws of Saltwater Country unsettle a white/Western level of consciousness that focuses on human mastery and possession that impedes racial justice, climate justice, multispecies and planetary justice. Like angry, hurt and frustrated Yirrkala bark painters who responded generously when they encountered the head of Baru, their ancestral crocodile, in a hessian bag at an illegal barramundi fishing camp in the 1990s (Buku-Larrngay Mulka Centre, 1999), sand art and video artworks are beautiful as well as generous experiments that respond to injustice through embodying an environmental ethics of responsibility. We see these experiments that offer spiritual gifts of connection to Saltwater Country as act of co-resistance to the violence of injustice. As Indigenous Potawatomi philosopher Whyte (2018) might say, perhaps this is one way of returning gifts from our Indigenous and Southern ancestors and becoming good ancestors ourselves.

Our experimental multisensory storying of Saltwater Country is a decolonial praxis that meshes photographs, videos, cultural knowledge, ceremony and rituals that affirm becomings with Saltwater Country and the possibilities for oceanic justice. Indigenous and migrant crosscurrents of saltwater thinking from the Indian and Southern Ocean comingle in this chapter. This is an oceanic turn that positions mobile sand grains as a temporary 'repository' (Agard-Jones, 2012, p 325) for affective, spiritual and elemental sensibilities or, as Yunkaporta (2019, p 17) would say, 'sand talks'. Our aim is not to romanticize these sensibilities that appear to give voice to the sand, wind, tides, waves, currents, wind and sea mammals. Michele is implicated in 'ancestral dystopias' that Indigenous peoples continue to endure (Whyte, 2018, p 225), but as a racialized brown settler, she cannot claim white innocence or saviourism like white settler allies.

The grounded, fluid and aerial vision presented in this chapter contrasts with mainstream geospatial ontologies that use scientific expertise as well as artificial intelligence to document and address the effects of climate change in coastal zones. The transitory sand artworks captured by drones, the production of videos produced using digital technologies as well as the ceremonies and rituals explored in this chapter are inspired by connections to sacred and kinetic forces. For example, the deep-time sand designs carved in Gunditjmara and Wadawurrung Sea Country captured by drone technologies fly with the wind and the energy of Bunjil, the creator spirit. The 'life force' of saltwater embodied in dreaming, healing and reinvigoration travels with electrical pulses through the internet and social media posts as well as art exhibitions, dance performances and intergenerational cultural workshops. Perhaps these ocean literacies and oceanic affects are embodied and digital performances that produce environmental stewardship or what Whyte (2013, p 117) calls 'systems of responsibilities' that can contribute to institutional change through a 'forward-looking framework of justice'.

References

Agard-Jones, V. (2012) 'What the sands remember', *GLG: A Journal of Lesbian and Gay Studies*, 18(2–3): 325–346.

Baldwin, A. (2016) 'Premediation and white affect: climate change and migration in critical perspective', *Transactions of the Institute of British Geographers*, 41: 78–90.

Bradley, J. with Yanyuwa Families (2010) *Singing Saltwater Country: Journey to the Songlines of Carpentaria*, Crows Nest, NSW: Allen & Unwin.

Brown, C. (2022). 'About 200 dead whales have been towed out to sea off Tasmania – and what happens next is a true marvel of nature', *The Conversation* [Online], 24 September, Available from: https://theconversation.com/about-200-dead-whales-have-been-towed-out-to-sea-off-tasmania-and-what-happens-next-is-a-true-marvel-of-nature-191340 [Accessed 4 January 2023].

Buku-Larrngay Mulka Centre (1999) *Saltwater: Yirrkala Bark Paintings of Sea Country, Recognising Indigenous Sea Rights*, Yirrkala: Jennifer Isaacs Publishing.

Bundle, Y., Rushton, J. and Wade, L. (2022) 'Citizens declare protection of whale songline country', *Friends of the Earth Australia*, 22 August, Available from: https://www.foe.org.au/citizens_declare_protection_of_whale_songline_country [Accessed 4 January 2023].

Clayton, R. (2022) 'More than 2,000 submissions made on proposed Viva Energy LNG terminal in Geelong', *ABC News* [Online], 9 May, Available from: https://www.abc.net.au/news/2022-05-09/viva-energy-lng-natural-gas-import-terminal-corio-bay-opposition/101047076 [Accessed 11 June 2022].

Escobar, A. (2020) *Pluriversal Politics: The Real and the Possible*, Durham, NC: Duke University Press.

Gay'wu Group of Women (2019*) Song Spirals: Sharing Women's Wisdom of Country through Songlines*, Crows Nest, NSW: Allen & Unwin.

Gergan, M., Smith, S. and Vasudevan, P. (2018) 'Earth beyond repair: race and apocalypse in collective imagination', *Environment and Planning D: Society and Space*, 1–15, DOI: 10.1177/0263775818756079.

Gilmore, R. (2022) 'Foreword: Spectacles', in R. Maynard and L.B. Simpson (eds) *Rehearsals for Living*, Chicago: Haymarket Books, pp 1–4.

Hartman, S. (2019) *Wayward Lives, Beautiful Experiments: Intimate Histories of Social Upheaval*, New York: W Norton and Co.

Hunter, L. and Lobo, M. (2022) *Climate Change and Oceanic Responsibilities: Listening and Dancing with Saltwater Country*. Video artwork presented at Crosscurrents of Oceanic Thinking, Association of American Geographers Annual Meeting, New York, 26 February.

Indigenous Peoples' Organisation-Australia (2021) 'Heal country, heal climate: priorities for climate and environment', Available from: https://www.ohchr.org/sites/default/files/2022-03/indigenous-peoples-organization2.pdf [Accessed 10 June 2021].

Ingersoll, K. (2016) *Waves of Knowing: A Seascape Epistemology*, Durham, NC: Duke University Press.

Lobo, M. (2014) 'Affective energies: sensory bodies on the beach in Darwin, Australia', *Emotion, Space and Society*, 12: 101–109.

Lobo, M., Alam, A. and Bandyopadhay, S. (2022) 'Tiger atmospheres and co-belonging in mangrove worlds', *Environment and Planning E: Nature and Space*, 1–22, DOI: 10.1177/25148486221079465.

Maynard, R. and Simpson L.B. (2022) *Rehearsals for Living*, Chicago: Haymarket Books.

Moreton-Robinson, A. (2015) *The White Possessive: Property, Power, and Indigenous Sovereignty*, Minneapolis: Minnesota University Press.

Okafor-Yarwood, I. (2022) 'Commentary: the more things change, the more they remain the same', *Political Geography*, 96: 102618.

Squire, R., Adey, P. and Jensen, R.B. (2022) 'Toward analog geographies: moving with and beyond enclosure', *GeoHumanities*, 8(2): 518–536, DOI: 10.1080/2373566X.2022.2108718.

Srivastava, S., Bose, S., Parthasarathy, D. and Mehta, L. (2022) 'Climate justice for whom? Understanding the vernaculars of climate action and justice in marginal environments of India', *Institute of Development Studies Bulletin*, 53(4, Available from: https://bulletin.ids.ac.uk/index.php/idsbo/article/view/3184 [Accessed 21 December 2022].

Tuck, E. and Wayne Yang, K. (2012) 'Decolonization is not a metaphor', *Decolonization Indigeneity, Education & Society*, 1(1): 1–40.

Wadawurrung Traditional Owners Aboriginal Corporation (2020) *Paleert Tjaara Dja, Let's Make Country Good Together, 2020–2030*, Geelong: Wadawurrung Traditional Owners Aboriginal Corporation.

Warren V.E., Constantine R., Noad M., Garrigue C. and Garland E.C. (2020) 'Migratory insights from singing humpback whales recorded around central New Zealand', *Royal Society of Open Science*, 7: 201084, DOI: http://dx.doi.org/10.1098/rsos.201084.

Whyte, K.P. (2013) 'Justice forward: tribes, climate adaptation and responsibility', *Climatic Change*, 120: 117–130.

Whyte, K.P. (2018) 'Indigenous science (fiction) for the Anthropocene: ancestral dystopias and fantasies of climate change crises', *Environment and Planning E: Nature and Space*, 1(1–2): 224–242.

Yunkaporta, T. (2019) *Sand Talk: How Indigenous Thinking Can Save the World*, Melbourne: Text Publishing.

INTERSTICE 1

Saturated Strands of (In/Re)Surgent Solidarity

Yin Paradies

Part I of this edited collection is entitled *Solidarity as Responsibility, Resurgence and Regeneration*. I will attend to three strands which I sense are braided throughout the chapters in this section: justice, solidarity and the aqueous. Environmental, ecological, climate, water, planetary, relational, Indigenous and multispecies justice are splashed throughout these chapters. For example, Baird (Chapter 3) reminds us that environmental justice is human-centric, while ecological justice is entangled with the more-than-human. Commonly, the concept of justice is enmeshed with laws and the legal/justice system in Western societies. It is evident that Indigenous lore (literally, a furrow or track contrasted with 'law', which means 'something laid down, that which is fixed or set' (*Online Etymology Dictionary*, n.d.) of Country applies equally to all, whether Indigenous or non-Indigenous. Country here refers to land, water, air, people, animals, plants, stories, songs, feelings and so on as they exist in merging waves of place-time (Gay'wu Group of Women, 2019). Manifestly, we all need to live once again 'in ways that are mutually respectful of all more-than-human relatives' (Poelina et al, Chapter 2); through Indigenous lore as nonoppressive guidelines, codes of conduct or nonconfining structures of place-based kinship formulated by creator beings, and continuously regenerated through everyday practices of renewal. One can only experience these Indigenous lifeways through the ongoing enactment of their constitutive relationships (Campion et al, 2023). Such a life cannot be abstracted, quantified, rationalized, determined, instrumentalized or standardized, without simultaneously extinguishing it.

It has been asserted that 'decolonization is not a metonym for social justice' (Tuck and Wayne Yang, 2012, p 21), with true decolonization requiring the return of land and/or settler-invaders going home. However, many of those to whom land could have been returned have already been murdered by invaders, while settler subjectivities (together with a profound loss of traditional knowledges) have become deeply embedded in most remaining descendants of precolonial First Peoples. In practice, 'decolonial crosscurrents of dreaming, reinvigoration and healing' beyond the risks of domesticated tropes of justice that recentre 'white innocence and white futures' (Hunter and Lobo, Chapter 5) are a clarion call to hospice settler subjectivities and become 'Indigenous to place' by feeling deeply 'in relation with more-than-human worlds' (Poelina et al, Chapter 2). How can we attend to neglected Country (Paradies, 2020), heal existing, and form new, authentic relationships, as if our lives (and those of future generations) depend on it, because they do (Kimmerer, 2011)? The potential for this is, perhaps, what Poelina et al describe in 'socioecological reinvigoration *as* cultural resurgence' (Chapter 2).

Rather than Poelina et al's entreaty to 'move on or live as a nation', I would contend that decolonization beyond modernist notions of justice is a move to relinquish the 'superfluous evils' – forms of unbearable suffering that could be prevented, but that persist through either design or neglect (Ophir, 2005) – which constitute the 'cold monster' (Nietzsche, 1969) of the nation (state). As Yunkaporta asks (2019, p 267), when will the Australian people be beyond saying 'sorry' to Indigenous Peoples and start saying 'thank you' or even 'please'? What would it feel like to start seeding and nourishing the 'lives to be lived once the settler nation is gone' (Tuck and Wayne Yang, 2012, p 36) as incipient response-ability to resurgent regeneration?

What if 'justice' itself is a problematic modern(ist) concept based on notions of grievance, redress, detachment, dispassion, objectivity, impartiality, rationality, hierarchy, binary balancing of scales, abstractions, judgement, redress and punishment? What if justice gives us a seat at the table, but doesn't allow us to frolic outside in the playground (Akomolafe, 2021)? Is justice inevitably framed upon bones, grief, tears, demise, death, erasure and silence? Does justice know how to meet the ghosts (Akomolafe, 2021), wraiths, spectres, shades and hauntings of history?

Orthogonal to 'disembodied critical debates ... "led by Western scholars and elites"' (Hunter and Lobo, Chapter 5), can we (as a more-than-human creaturely collective) (re)form justice into something much more capacious that 'embod[ies] an environmental ethics of responsibility' (Hunter and Lobo, Chapter 5)? Could 'climate justice' become 'climate care' (Poelina et al, Chapter 2)? Perhaps justice can grow into earthly healing, synergistic responsibility, critical consciousness and self-determined realities, as a commitment to right relationships between humans, so-called 'nature' and the spiritual realm, encompassing ethical praxis such as seeking truth, harmony, balance and healing (Chioneso et al, 2020).

This brings us to the second strand amongst these chapters: solidarity considered as a 'need' (Poelina et al, Chapter 2), 'necessity' (Baird, Chapter 3), 'crucial' and 'undervalued' (Haldar et al, Chapter 4). Some have noted that 'solidarity ... [is] essential as we attempt to dismantle the death-machine' (Green Anarchy Collective, n.d.). However, along with justice, (false) solidarity as a practice has 'never been mutual' and has been 'an instrument of settler colonial assimilation ... skewed towards the pursuit of a settler colonial future' (Benally, 2021, p 12).

Poelina et al (Chapter 2) aptly note that instead of an 'earthly co-existence', we, as a modern society, are incessantly chasing the 'almighty dollar sign in the cloud' and, importantly, that 'crises of sustainability are crises of relationship' and hence crises of solidarity. How can we invite 'kinships of inextricable rapport' (Poelina et al, Chapter 2) with which to 'unsettle a white/western level of consciousness' (Hunter and Lobo, Chapter 5)? At the very least, to embrace radical new forms of more-than-human solidarity, we need to engage with Indigenous ways of knowing, doing and being that are vastly divergent, and largely incommensurate, with Western epistemologies, axiologies and ontologies. Indigenous perspectives are contextual, local and intermingled; innately ritualistic, ceremonial, reverential and teeming with grace; as well as inherently attentive, receptive, prefigurative, patterned and layered, in which the ends never justify the means and concepts/ideas stem from deeds/enactment rather than vice versa. By 'ritual', I mean rhythmic repetitive resonant vocal, musical and/or kinaesthetic communal animate enacted entrainment. This often involves relinquishment of agency and austerity/intensity (exertion, fasting, pain, piercing, burning, cold, heat, burial and so on) to re-fuse (with) the world; conjunct with the cosmos, cross the threshold from linear to eternal temporalities and enter trance states.

Against the prevailing and pervasive Eurocentric story of separation, rather than framed isolated insulated distinct things, Indigenous worlds are constituted by relationship, connectedness and the space between (Poelina et al, Chapter 2). In such worlds, reality can be (at times) experienced as a blended continuum/indivisible duration, in which there is no dichotomy between the material and immaterial and no essential difference between any of us and the rest of the universe, at any place or time (Allen, 2022). In other words, all forms are ever-always in a process of interdependent co-arising, with no-thing pre-existing the relations which comprise it (Escobar, 2019). Such existence-scapes have profound implications for the practice of solidarity as it emerges in community, participation, intimacy and sharing within a spontaneous emergent complex self-organizing cosmos. Indigenous realities are not only literal, logical, causal, rational and factual, but also mysterious, mystical, mythical, magical, metaphorical and melodious. In such realities, many modern ideas, contested but assumed to be universal, are often absent (or virtually so), perhaps including the notion of solidarity itself.

What if, due to inherent incommensurabilities between settler and native subjectivities, some solidarities are undesirable (Tuck and Yang, 2012)? Or perhaps, beyond suppressed solidarities, the 'work of foraging with solidaristic transgressions ... may per chance translate into feminist, anticolonial, anticapitalist world-making' (Vijay, 2023, p 428) in which 'our solidarity is projected out from our relationship with the Earth' (Benally, 2021, p 24). Hunter and Lobo (Chapter 5) yearn for 'a decolonial praxis of solidarity ... [that] unsettles the language of climate crises, *aqua nullius* and "white eco-apocalyptic futures"'. Certainly, an 'affirmative critique ... that galvanizes people to come together with enthusiasm' for 'everyday political activism ... attune[d] to the more-than-human life force' is required to counter 'the brutal racial logics of "human" mastery and white supremacy' (Hunter and Lobo, Chapter 5).

Poelina et al call for 'transformative change' that is 'solutionary' 'rather than revolutionary (Chapter 2). I would contend that any true transformation or metamorphosis follows a communal insurrectionary spiral rather than being about either another turn of the wheel (revolutionary) or set within a problem–solution paradigm. What we face are much more profound predicaments that we can, together, attend to, sense into, respond to and anticipate perturbations of, rather than 'fix' or 'solve' as such. Importantly, these predicaments call upon our collective solidaristic response-abilities rather than individual heroic efforts.

'The pathway forward demands being brave' (Poelina et al, Chapter 2), so very brave, vulnerable, courageous, and tenacious if we want to re-generate societal assemblages which preclude power-over, supremacy, superiority, domination (Poelina et al, Chapter 2) and resurrect cultural forms and flows that are non-exploitative and nonauthoritarian (Coulthard and Simpson, 2016). 'This is not an intellectual process, it is a heart process, a spiritual process, an embodied process' (Poelina et al, Chapter 2). Rather than 'restoring agency to ... more-than-human beings' (Poelina et al, Chapter 2), this is a process of re-membering the sentient and sapience which has always been, among all that abounds. This is patterned perception of plagency in place, an appeal to embrace 'constellations of co-resistance', 'beautiful experiments' and 'responsibility for an earth otherwise' while staying with the comingled joy and pain 'of ... listening and dancing with ... embodied, visual, affective, kinaesthetic and spiritual ... energies, songs, feelings, and stories' (Hunter and Lobo, Chapter 5).

On this journey, we are courageously challenged to meet our 'complicity with colonization through our daily lives – in supermarkets, kitchens and energy systems' that drives us towards civilizational collapse (Poelina et al, Chapter 2). We are also asked to face the fact that Indigenous cultures are deeply (relationally) individualistic, with far more freedom, peculiarity, eccentricity and 'intimacy among relatives of infinite diversity' (Bird-David, 2017, p 223) than any modern nation state allows or could ever support. Being held within communal distributed networked care and compassion (meaning literally 'to suffer with') in fiercely egalitarian contexts enables 'dividual' autonomy to thrive and flourish. Here, Strathern's (1988) notion of a 'dividual' (a person constituted by relationships) helps us perceive the contrast with an 'individual' (a single bounded integrated entity).

This combination of autonomy and (inter)dependency in Indigenous cultures evokes relationships that are strongly founded on trust, itself founded on each being's response-ability (ability to respond) with integrity. To trust another being is to act with that being in mind, in the hope and expectation that they will do likewise, by responding in ways nourishing to you. However, on no account should one attempt to force a response by placing another person under obligation or compulsion. To do so would represent a betrayal of the trust you have placed in them. This would be tantamount to a renunciation of the relationship. Trust is founded on a respect for the (relational) autonomy of the other on whom one depends. Trust is conditional on situational leaders respecting their followers' autonomy. Should a leader, at any stage, seek to dominate a follower, whether by threat or command, the follower, feeling their trust betrayed, is likely to take their loyalty elsewhere. In this way, power, in Indigenous cultures, works via attraction, not coercion, and the slightest tip in the balance towards domination will cause trust to be lost (Lee and Daly, 2012).

The third strand that is braided through these chapters is a focus on the oceanic and water(ways) along with sacred entities such as the water snake (Poelina et al, Chapter 2). As Baird (Chapter 3) notes, 'freshwater is increasingly referred to as "liquid gold" or "blue gold"', understood as a vital resource for all life. This level of comprehension allows us to recognize 'the operation of the water markets ... as unfair to the critical water needs of local farmers' (Baird, Chapter 3) enacting environmental racism, water colonialism, ecocide and state-facilitated water theft.

In the Sundarban region of India and Bangladesh, which experiences significant water shortages, deprivation and conflict-inducing water commerce, Haldar et al (Chapter 4) contend that 'earth system governance and planetary stewardship', adaptive technologies, behaviour change, equitable meaningful stakeholder participation, reduced corruption/malpractice as well as 'a holistic, well-planned and long-term strategy' are all 'crucial for healthy water systems'. However, acknowledgement of such fluid injustices and governance/planning requirements does not help us eclipse 'the commodification, atomization and segmentation' of water (Baird, Chapter 3). How can we learn to encounter water beings as free-flowing entities (Baird, Chapter 3), spirited life that ebbs and flows as our kin (Mehltretter et al, 2023)? Also through the lens of encounters with the Ganges in the Sundarban ecosystem, da Cunha (2018) queries the very notion of rivers, eliciting instead an environs of ubiquitous wetness, in which water permeates air, soil, seas, fields, buildings, plants, animals and minerals, while Schrei (2021) highlights the many traditions that have viewed consciousness as an ocean, gesturing towards our human conscious processes which occur within a matrix of water; a reciprocity (meaning literally 'back and forth'), like a pulse of life beyond the 'terra firma–aqua fluxus divide' (da Cunha, 2018).

In his research, Baird finds some cause for optimism in farmers turning to ecological justice, 'reject[ing] the binary logic of "human" and "nature" in recogniz[ing] both the instrumental and intrinsic value of fresh water', while 'strengthen[ing] their social, economic and spiritual connections with the ecosystems in which they see themselves as entangled' (Baird, Chapter 3). Hunter and Lobo (Chapter 5) also conjure hope via 'the beautiful and generous gifts' of 'multisensory, visceral and immersive encounters' with the 'spiritual touch' of littoral oceanic affects, atmospheres and intensities. Poelina et al (Chapter 2) highlight the importance of 'living place-based Indigenous knowledge existing in a nonlinear time scale of past-presence-emergence'. Such emergent, eternal, spiral, spatial, slushy, textured, everywhen temporalities are inimical to the settler-colonial timescapes that may one day be redundant (again) in a cosmic kismet of the future-present. 'By looking back and forth (future-past), we can walk in the present-future' (Belanger et al, 2023), perceiving and consciously living the symbiotic with-ness of life nested across the possibilities of possibilities that ripple and resonate outward from each and every intermingled moment-location across thrumming waves of place-time.

'Stewardship of Country is enmeshed in a *relational* ethic of care and love' through 'kincentric ecology' and 'a collective … rather than … egotistical sense of self' (Poelina et al, Chapter 2). Through living this understanding in embodied and environed ways, we become trusted ancestors contributing to the unfurling of sublime wonder in a symphony of co-becoming. As Poelina et al artfully express, 'ancient wisdom transforms the banal into the beautiful, the significant and the sacred' (Chapter 2). The quotidian nature of the holy and divine are foundational to Indigenous experiences (Paradies and Joyce 2024) in which nothing is complete, perfect or enduring, but all is alive, sentient, profoundly relational and deeply sacred, enfolded in a cosmos braided from the finest filaments of blazing consciousness.

References

Akomolafe, B. (2021) 'We will dance with mountains' [Online course], Available from: https://www.bayoakomolafe.net/offerings [Accessed 12 March 2024].

Allen, D. (2022) *Ad Radicem: To the Root!*, England: Expressive Egg Books.

Belanger, P., Jafari, G. and Escudero, P. (2023) *A Botany of Violence: Across 528 Years of Resistance and Resurgence*, California, U.S.: Goff Books and ORO Editions.

Benally, Klee Ya'iishjááshch'ilí (2021) 'Unknowable: against an Indigenous anarchist theory', in *Black Seed: Not on Any Map*, June.

Bird-David, N. (2017) *Us, Relatives: Scaling and Plural Life in a Forager World*, Oakland: University of California Press.

Campion, O.T. et al (2023) 'Balpara: a practical approach to working with ontological difference in Indigenous land and sea management', *Society & Natural Resources: An International Journal*, DOI: 10.1080/08941920.2023.2199690.

Chioneso, N.A., Hunter, C.D., Gobin, R.L. McNeil Smith, S., Mendenhall R. and Neville, H.A. (2020) 'Community healing and resistance through storytelling: a framework to address racial trauma in Africana communities', *Journal of Black Psychology*, 46(2–3): 95–121.

Coulthard, G. and Simpson, L. (2016) 'Grounded normativity/place-based solidarity', *American Quarterly*, 68(2): 249–255.

Da Cunha, D. (2018) *The Invention of Rivers: Alexander's Eye and Ganga's Descent*, Philadelphia: University of Pennsylvania Press.

Escobar, A. (2019) 'Habitability and design: radical interdependence and re-earthing of cities', *Geoforum*, 101: 132–140.

Green Anarchy Collective (n.d.) *What Is Green Anarchy? An Introduction to Anti-Civilization Thought*.

Gay'wu Group of Women (2019) *Songspirals: Sharing Women's Wisdom of Country through Songlines*, Crows Nest, NSW: Allen & Unwin.

Kimmerer, R. (2011) 'Restoration and reciprocity: the contributions of traditional ecological knowledge', in D. Egan, E.E. Hjerpe and J. Abrams (eds) *Human Dimensions of Ecological Restoration. Society for Ecological Restoration*, Washington DC: Island Press, pp 257–276.

Lee, R. and Daly, R. (eds) (2012) *The Cambridge Encyclopedia of Hunters and Gatherers*, Cambridge: Cambridge University Press.

Mehltretter, S., Longboat, S., Luby, B. and Bradford, A. (2023) *Indigenous and Western Knowledge: Bringing Diverse Understandings of Water Together in Practice*, Technical Report for the Global Commission on the Economics of Water. Convened by the Government of the Netherlands, facilitated by the Organisation for Economic Co-operation and Development (OECD).

Nietzsche, F. (1969) *Thus Spoke Zarathustra*, R.J. Hollingdale (trans), London: Penguin.

Online Etymology Dictionary (n.d.) 'Law', Available from: https://www.etymonline.com/search?q=law [Accessed 12 March 2024].

Ophir, A. (2005) *The Order of Evils: Toward an Ontology of Morals*, R. Mazali and C. Havi (trans), New York: Zone Books.

Paradies, Y. (2020) 'Unsettling truths: modernity, (de-)coloniality and Indigenous futures', *Postcolonial Studies*, 23(4): 438–456.

Paradies, Y. and Joyce, C. (2024) 'From Esotericism to Embodied Ritual: Care for Country as Religious Experience', Religions, 15(2): 182.

Schrei, J.M. (2021) 'The many voices of water, part 1: oceans of melancholy and bliss', *The Emerald Podcast*, Emerald Publishing, Available from: https://podcasts.apple.com/us/podcast/many-voices-water-part-1-oceans-melancholy-bliss/id1465445746?i=1000507825410 (Accessed 29 March 2024).

Strathern, M. (1988) *The Gender of the Gift: Problems with Women and Problems with Society in Melanesia*, Berkeley: University of California Press.

Tuck, E. and Wayne Yang, K. (2012) 'Decolonization is not a metaphor', *Decolonization: Indigeneity, Education and Society*, 1(1): 1–40.

Vijay, D. (2023) 'Settled knowledge practices, truncated imaginations', *Organization*, 30(2): 424–429.

Yunkaporta, T. (2019) *Sand Talk: How Indigenous Thinking Can Save the World.* Melbourne: Text Publishing.

PART II

Solidarity without Borders

6

Asserting Indigenous Self-Determination and Climate Justice Through Resisting Coal: A Global North–South Comparison

Ruchira Talukdar

Introduction

> We exist as people of our land and waters, and all things on and in them – plants and animals – have special meaning to us and tell us who we are ... If they are destroyed, we will become nothing ... we have not consented to the development of the Carmichael coalmine or any other proposed mine on our traditional lands.
> Wangan and Jagalingou Traditional Owners, Australia,
> *Submission to the Special Rapporteur on Indigenous Peoples*, 2015

This chapter aims to understand how Indigenous resistances to extractive fossil fuel capitalism mobilize based on self-determination through the Free, Prior and Informed Consent (FPIC) principle. Such struggles are at once about redressing historic wrongs of land dispossession from colonization, its continuation through ongoing dispossession, and securing just climate futures for Indigenous communities who are among the worlds' most vulnerable to climate change. To gain a global perspective on the legal and political possibilities for Indigenous land struggles against fossil fuel projects to assert the FPIC, this chapter compares two Indigenous struggles against coal mining in Australia and India: two coal-dependent democracies representative of the Global North, settler colonial (Australia) and Global South, postcolonial (India) contexts.

I ask the following questions: how is the FPIC being mobilized by Indigenous anti-fossil fuel resistances to achieve land and climate justice in the Global North and South, in settler colonial and postcolonial contexts? And how can climate activism forge solidarities with these struggles for past, present and future justice? Through this comparative analysis, I make recommendations for how climate activist solidarities can be forged.

I compare the struggle of the Wangan and Jagalingou (W&J) traditional owners of central Queensland, Australia, and their collaboration with the mainstream climate movement, to resist the Carmichael coalmine, with the struggle of Mahan forest dwellers in Madhya Pradesh, India, to stop a coalmine in collaboration with the environmental group Greenpeace. Both resistances intersect matters of Indigenous land justice and climate justice, as I outline in the following paragraphs.

Owing to the continuation of colonial dispossession through attacks on Indigenous land, the idea of Indigenous justice tends to be centred on land and sovereignty as redressal for historic wrongs (Norman, 2015). Attacks on Indigenous lands through a nexus of state and corporation can include extractive projects, including for the mining of fossil fuels – coal, oil and gas – that are primary drivers of climate change. But they can also include renewable, 'clean energy' and 'climate solutions' that proceed without Indigenous consent. Far beyond climate activists' singular framing of fossil fuels as the 'problem' and renewable energy as the 'solution' to climate change, Indigenous climate justice considers past, present and future human and ecological harm. Indigenous climate justice is fundamentally linked to sovereignty and self-determination based on land, and rights over land (Birch, 2018; Lyons et al, 2021).

The principle of FPIC entails the right for Indigenous people to be informed, to participate in decision making and to give or withhold consent. Indigenous communities consider the FPIC a central tool to arrest expropriation of lands, resources, knowledge and culture (Carino, 2005). This makes FPIC an essential tool for Indigenous land justice and climate justice. Through the legal requirement of FPIC, consent can be given or withheld, and nonconsensual extraction, development or resource-use can be challenged (Perera, 2016). FPIC is a key instrument for Indigenous land rights in international law, in human rights instruments such as the United Nations Declaration of the Rights of Indigenous People (UNDRIP) and in environmental conventions. It is a substantial requirement, particularly in the case of mining and other extractions. In international law, FPIC is interlinked with the fundamental Indigenous human right to self-determination (Giacomini, 2022).

National and subnational land and forest rights regimes are often the only legal means available for Indigenous communities to redress historic injustices and ensure future justice through securing titles and tenures over lands and forests. In the Global North and South, Indigenous resistance to extraction has mobilized around provisions of self-determination in the land-rights-based legislation of states. But such legislation might not necessarily uphold the international standard of FPIC enshrined in international instruments. Further, even when Indigenous communities have secured their legal rights over land and to consent, they often lack the power to meaningfully make decisions, owing to their power inequality with the state-corporate machinery as socially vulnerable communities (Chowdhury and Aga, 2020). Provisions for Indigenous self-determination as enshrined in states' legislation, whether and how this reflects the international standard of FPIC, as well as the practices of the state-corporate machinery, all play a pivotal role in enabling (or withholding) Indigenous climate justice (Talukdar, 2021).

The W&J and the Mahan Indigenous struggles are both organized around national land rights legislation that governs consent and participation of Indigenous people in making decisions about extractive projects on their lands. Both assert their right to FPIC – their self-determination – by refusing consent to coalmining on their lands. The two forms of resistance began in 2011. At Mahan, although coalmining was stopped in 2014, challenges to securing Indigenous forest rights have continued. The W&J's struggle is sustained in various forms before and after coalmining began at Carmichael in 2020.

Drawing on multilocational ethnography conducted in Australia and India between 2017 and 2019, I compare critical moments of the W&J struggle (2012–2017), with that of the Mahan struggle (2012–2015), focusing on how the two campaigned for self-determination. The first section of this chapter outlines my approach to fieldwork, interviews and analyses of findings. I outline how my positionality as a scholar activist influenced my analysis of the cases. The second section establishes the connection between land rights, climate justice, self-determination and FPIC, as framed in international human rights instruments. The third and fourth sections discuss the Australian and Indian cases respectively. These sections outline the power context produced by the state-corporate nexus for mining, in which the respective land rights regimes operate. The sections also discuss the Indigenous self-determination campaigns. The fifth, drawing on the third and fourth sections, delineates differences and similarities in the two contexts for Indigenous self-determination. The final section recommends how activists and scholars can frame solidarity with Indigenous anti-fossil fuel struggles for self-determination.

Approach and methods

I draw on my PhD research, conducted between 2017 and 2019, for a comparative ethnography of anti-coal resistances in Australia and India. The multilocational ethnography included three field visits to Singrauli – in villages across Mahan – and two in central Queensland – close to the Carmichael mine site. It included participant observation at environmental nongovernmental organization (ENGO) meetings and public hearings in India (in Mahan and various urban locations), and climate gatherings at locations across Australia where the W&J were also present. I conducted 25 interviews in Australia. The relevant interviews for this chapter are with two W&J spokespersons, and the campaign manager for W&J's struggle against coal mining (a non-Indigenous person). I conducted 25 interviews in India. The relevant interviews for this chapter were with five male and one female spokespersons of the Mahan resistance. All Indigenous spokespersons consented to being interviewed and recorded. I limited my interviews to designated spokespersons for the W&J and Mahan campaigns to mitigate the risk of harm to community members that could arise from discussing sensitive issues. The fieldwork and interviews were analysed and thematically sorted into comparable categories. Documents analysed for the PhD research between 2011 and 2019 that are relevant for this chapter include news articles on the W&J and Mahan movements, press releases and reports issued by the two movements.

I also draw on my insights as a climate justice campaigner of nearly two decades across Australia and India. This research and my experiences have made me realize the need for mainstream climate activism to forge radical solidarities with Indigenous struggles for land and self-determination. Such solidarities would keep questions of Indigenous futures at their heart. And they would move away from an approach of what Vincent and Neale (2016) call 'tactical alliances': unstable, short-lived and forged with a singular focus on stopping mining. In this chapter I argue that the changed focus of global climate activism, predicated on forming solidarities with local struggles, necessitates a deep rethinking of the nature of alliance making across Global North-South intersectional differences.

The term 'intersectionality' was first used in critical race theorization and was coined by Kimberlé Williams Crenshaw (2017). This chapter uses intersectionality as conceptualized in climate justice praxis. Because 'we are all in this together', the climate era can create an inclusive ecological justice and an intersectional politics across various environmentalisms that is less like exclusionary (Northern) environmentalism and more like human rights, since 'what we are fighting for is each other' (Stephenson, 2015, p xv). The task for climate activists is to overcome the various disconnections and connect the various movements. For this, ecocentric Northern environmentalism will have to decolonize and establish solidarity with the vision of movements that are fighting to redress historic injustices (Klein, 2016; see also Chapter 2 in this volume).

Yet these possibilities for intersectionality and the notion of climate justice as a common frame have emerged from the Global North. This chapter therefore also extends the significance of intersectionality for climate justice to be cognisant and inclusive of Northern-Southern contextual differences, while supporting global Indigenous climate justice outcomes.

Land-based self-determination in international instruments: settler colonial and postcolonial contexts

Settler colonial pertains to a sociopolitical context in which the colonizers remained, as in the case of Australia. This contrasts with a postcolonial context in which a state has been formally decolonized through the act of colonizers leaving, as is the case in India. Mining and other extractive projects on Indigenous lands perpetuate the experience of colonization for Indigenous peoples, bringing different political and ethical implications to bear on them in the two contexts.

As an industrialized country, Australia belongs to the Global North, and India, as a developing country, to the Global South. Under the United Nations' framework for climate change negotiations (UNFCCC), the terms signify how countries politicize their climate responsibilities, with developing countries like India holding developed countries historically responsible. These categories are relevant for the multiscalar case study in this chapter.

Indigenous peoples comprise 1 per cent of the world's population, one third of the world's poor and hold one fifth of the world's land area, with much of the world's biodiversity and deposits of minerals and fossil fuels (UN, 2021). Climate change is felt most directly by people on the land who are closest to its impacts; this makes Indigenous peoples and local communities (IPLCs) among the most affected. IPCLs are also directly affected by the extraction of fossil fuels – coal, oil and gas – that are the predominant drivers of climate change. Indigenous people defending their land and human rights also face systematic violence from states and corporations (Giacomini, 2022). The connection between securing land tenures for Indigenous climate justice and achieving better land and forest management for climate action through Indigenous governance is increasingly recognized (Agarwal and Dash, 2022).

Given the political economy of colonial-capitalist extraction, Indigenous people are likely to be revictimized by land use changes in transitions to renewables and climate solutions that ignore Indigenous voices while promoting market-driven technologies (Mookerjea, 2019; Porter et al, 2020). Energy transitions need to acknowledge historic wrongs of dispossession of Indigenous people, rupturing land-based relations and its continuation today. Indigenous land rights are understood as a way of initiating such an acknowledgement (Whyte, 2021).

A key challenge for Indigenous land rights and consequently Indigenous climate justice lies in the low levels of recognition of the former: although collectively Indigenous people and local communities manage over 50 per cent of the world's lands, their legally recognized rights comprise just over 10 per cent (RRI, 2017). This situation makes international human rights instruments such as the UNDRIP crucial for Indigenous land and climate justice – in particular, how these instruments enshrine rights and determination for Indigenous people through land, and how international standards are upheld in national (or subnational) laws and practices of states.

The UNDRIP, adopted by the UN General Assembly in 2007, includes the right to self-determination under Article 3, which states that 'by virtue of that right they freely pursue their economic, social and cultural development'. Article 26 qualifies self-determination as a remedial political right of distinct peoples and nations with a history of dispossession. The UNDRIP details situations where Indigenous people exercise self-determination through FPIC under various articles including Article 27, which pertains to the exploitations of lands, territories and natural resources. Given the nonbinding effect of the UNDRIP on ratifying states, whether self-determination is enshrined in national (or subnational) laws, and how state practices enable Indigenous self-determination have proven to be decisive factors in terms of whether Indigenous movements have been able to assert their human rights (Talukdar, 2021b).

Across the next three sections I compare legal regimes under Australia's Native Title Act 1993 (NTA) and India's Forest Rights Act 2006 (FRA), and two case study experiences of Indigenous anti-mining struggles with these laws in both countries.

Politics and mobilization for self-determination in Australia

Land rights: politics, limitations and benefits of native title
Since the 1967 referendum leading to the amendment of the Constitution to include Aboriginal Australians in the census, the ability of the Commonwealth to legislate on Indigenous affairs has impacted on the capacity to address the injustices of colonial dispossession (Altman, 2009). A historic strike by Aboriginal stockmen in 1966 that began with families walking off the Wave Hill pastoral station in the Northern Territory ultimately triggered a movement for Indigenous land rights. Yet the mining industry's influence over the state determined the extent of rights granted under land rights laws eventually passed. The politics of land rights sets an essential context for today's Indigenous struggles with fossil fuel centred on self-determination, particularly through Australia's NTA.

In the 1992 *Mabo* decision, the High Court overturned Australia's founding legal fiction of *terra nullius*, ruling that to deny Indigenous land rights contravened racial equality guaranteed under Australia's Racial Discrimination Act 1975 (RDA). But the *Mabo* decision recognized an extremely limited form of native title, granting limited occupation not even equivalent to a standard lease (see *Mabo* 1992). The NTA was passed following the Mabo decision and the Australian Parliament passed into law the limited framing of native title.[1]

If returning Indigenous lands was intended to redress dispossession caused by Australia's colonial project, the native title apparatus ensured mining interests remained empowered over informed Indigenous consent, capitalizing on divisions in the community (Bebbington et al, 2008; Coyne, 2017). Processes of agreement making, negotiation, and dispute resolution, between mining companies and native title groups make this evident. Native title groups have the right to negotiate with resource developers within six months of notification of a proposed project, after which the matter requires arbitration in the Native Title Tribunal (NTA) (Altman, 2012). During such arbitrations, groups have often been pressured to settle agreements or risk their native title being extinguished by the state without receiving any benefits from the company. Agreement-making processes between communities and mining companies are often characterized by a lack of transparency on part of companies, and confusion, criticism and dissent on the part of native title claimants (Norman, 2016).

Returning land and granting rights towards sovereignty are essential reparations for historic dispossessions and freedom from present colonial realities in a settler colonial society (Gilbert, 1994). But the NTA fell short of these actions, especially by withholding Indigenous groups' right to refuse consent over development on their lands. It fell short of the UNDRIP and other international instruments that emphasized free, prior and informed consent. Turner and Rojeck highlight a 'frequent tension between national systems of rights and international human rights' (2001, p 127). Owing to the NTA's limitations, Indigenous Australians have turned to the UNDRIP as a more accurate articulation of self-determination and land rights (Short, 2007). Struggles that resist fossil fuel extraction based on self-determination must negotiate this gap between the FPIC's intent (as enshrined in the UNDRIP) and its application (through the NTA).

Fossil fuel extraction from Indigenous lands increased significantly during Australia's latest minerals boom that began in the early 2000s and lasted until approximately 2015. The presence of the NTA shifted mining companies away from 'bareknuckle' racism in dealings with remote Aboriginal communities, demonstrated in earlier booms when mining occurred with no negotiations or benefits to Aboriginal people (Langton, 2012). Since the NTA's institution, native title has been conferred on more than 32 per cent of the Australian continent (AIATSIS, 2022).

'No means no': the Wangan and Jagalingou's challenge to native title

In 2004, the W&J traditional owners' native title claim was registered over 30,000 square kilometres in the Galilee Basin in Central Queensland. Five years later, the Queensland government began implementing its vision to develop the Galilee Basin for coal. The W&J's claim area includes lands that came under mining leases for the Adani Enterprises-owned Carmichael coalmine.

Wangan and Jagalingou's campaign and collaborations

Negotiations between Adani and the W&J about consent for mining, and an Indigenous land use agreement (ILUA) commenced in 2010. The W&J contended that Adani negotiated in bad faith during the stipulated six-month period by taking advantage of the coercive power of the native title system. During my interviews with W&J spokespersons, they identified these bad faith negotiations and Adani's below-par compensation offer[2] as important in their decision to fight the coalmine. They refused consent, becoming the first native title group in Australia to give an outright no to mining (Lyons et al, 2017a). The ILUA was struck down on three separate occasions at bona fide meetings of the W&J native title claim group (Lyons et al, 2017a).

Adani brought proceedings before the National Native Title Tribunal after each of the two rounds of failed negotiations in 2013 and 2015. Both times, the Tribunal ruled that the Queensland government could grant mining leases under the NTA (W&J, 2015). Queensland issued all the mining leases to Adani despite the W&J's lack of consent (Australian Associated Press, 2016). The mining leases and the Tribunal's determinations were based on what the W&J called a 'sham agreement' and a 'fake meeting to manufacture consent' (Lyons, 2017b), which I will discuss in the next subsection. The W&J's experience during the negotiations revealed a constant prioritization of mining and settler-state agendas over meaningful consent for Indigenous people. In 2015 the W&J Family Council withdrew from Tribunal proceedings, stating that:

> These proceedings and the legislation under which they are held do not advance our right to live in freedom, peace and security as distinct peoples with our own cultural values ... We are running our campaign based on the singular act of self-determination and our right to say 'no' as the Traditional Owners and custodians of our ancestral lands where our ancestors still reside. While the legal system might weigh against us – when we say No, we mean No! (W&J, 2015, para 21)

Next, the W&J Family Council mounted legal challenges and a public and international appeal for its right to FPIC as per the United Nations Declaration of the Rights of Indigenous Peoples under the campaign slogan 'Adani, No Means No' (W&J, 2015). Challenging the native title system became a focus of its resistance:

We are taking on the racist legacy of the Native Title Act and its failure to measure up to international laws that declare the rights of Indigenous people. We have sent a complaint through the UN Rapporteur on Indigenous Rights complaining about what they are doing to Indigenous people in this country. (Burragubba, 2016, para 12)

The W&J built solidarities with Indigenous movements fighting fossil fuels – with the Athabasca people in Canada fighting the Keystone Pipeline and Standing Rock Sioux and Chickaloon Village Traditional Council in Turtle Island in the US against the Dakota Access Pipeline (W&J, 2018). While reiterating the distinctiveness of their struggle, the W&J also strategically collaborated with the mainstream mobilization against the Carmichael coalmine in Australia, 'Stop Adani'. They toured international banks and investors across the US and Europe in 2015, in collaboration with Stop Adani, to convince international funders not to invest in the Carmichael coalmine (Market Forces, 2015):

The Wangan and Jagalingou people have joined with the environmental movement but we are running our own campaign, based on the singular act of self-determination and our right to say 'no' as the Traditional Owners and custodians of our ancestral lands … In the end we might have to sacrifice more than the others. (W&J Spokesperson Adrian Burrugabba at the 2016 'Beyond Coal and Gas Summit' near Newcastle in New South Wales, from summit notes, 4 April)

State and corporation manufacturing Indigenous consent

Adani claimed that unanimous agreement was achieved from W&J traditional owners in a community meeting it organized in April 2016 (Wagner, 2017). Adani organized the meeting in the coastal town of Maryborough, instead of the W&J's lands, where it was allegedly easier to 'rent-a-crowd' from nearby Aboriginal communities in Cherbourg and Woorabinda. The company paid for the transport and accommodation of 341 attendees, hundreds more than the size of any previous W&J meetings (Carey, 2019). The meeting was effectively stacked with attendees who did not have the right to vote on the Adani ILUA (Brigg, 2018). Although mining companies customarily cover meeting expenses, the process allows the corporations to offer inducements and consequently influence meeting outcomes.

In December 2016 the W&J Family Council legally challenged Adani's 'sham' agreement from the April 2016 meeting, alleging that a majority vote was obtained from a 'rent-a-crowd' Indigenous gathering, who had never before identified as W&J people (W&J, 2016). But the court ruled against them, indicating how the laws are stacked to favour mining corporations (Robertson and Sigato, 2018).

Exposing the limits of native title
While visiting Australia in September 2016, the UN Special Rapporteur on Human Rights Defender singled out the Carmichael coalmine as a case of poor Indigenous consultation, noting that since ILUA consultations had not been conducted in the spirit of FPIC principles under the UNDRIP, allowing the project to violate Australia's international obligations (Frost, 2016, para 7). But even repeated interventions from the UN could not stop the state from paving the way for Carmichael without Indigenous consent. The federal government even amended the NTA in a manner described as 'completely disrespectful' to Aboriginal people to facilitate the mine (*SBS News*, 2017).[3]

Amendments to the NTA and unfavourable rulings in legal challenges exhausted pathways for the W&J's challenge to native title. Although 'No Means No' could not stop the Carmichael mine from proceeding without Indigenous consent, the campaign gained recognition as a leading Indigenous rights struggle that exposed the limitations of Australian native title (Lyons, 2019).

Politics and mobilization for self-determination in India

Forest rights: politics, possibility and lost opportunity
Forests cover over 40 million hectares of India's landmass and support over 300 million livelihoods (FAO, 2017). Large sections of India's forests fall under the Fifth and Sixth Schedules of the Constitution. Scheduled Areas contain large populations of Indigenous tribal and Adivasi Indians[4] living in and depending on forests. The Fifth Schedule designates Adivasi-majority forested areas in central India in order to provide special protection for their lands. Central India has served as a site of conflict since British colonization; between the sovereign rights of Adivasis over *Jal, Jangal, Jameen* (water, forests and land) and extractive capitalism. Colonial laws such as the Forest Act 1878 were responsible for displacing thousands of forest-dwelling Adivasis through the creation of state-owned Reserved Forests. The colonial nature of extraction continued even after independence in 1947, with Indian governments taking over Adivasi lands in the same vein as their colonial counterpart (Ramesh and Khan, 2015).

Coalmining to fuel India's post-independence development became the chief agent of disruption for central India's forest communities (Lahiri-Dutt, 2016). Later on, as India's economy privatized (between 1994 and 2009) and mining for coal and other minerals rose by 75 per cent, nearly all proposed coalmines were allocated in central India and significant proportions in Fifth Schedule Areas (Shrivastava and Kothari, 2012). Against this political backdrop, the Forest Rights Act was introduced in 2006 as a national response to long grassroots movements for Adivasi forest rights – converged under the umbrella of the 'Campaign for Survival and Dignity' – at a time when India was witnessing increasing conflicts from increased mining (Kumar and Kerr, 2012; Sigamany, 2022).

The Scheduled Tribes and Other Traditional Forest Dwellers (Recognition of Forest Rights) Act 2006 (hereinafter FRA) aimed to redress colonial injustices by giving forest-dwelling tribal, Adivasi and non-Indigenous communities a say in how the state should deal with their forests (Ramesh and Khan, 2015). The preamble acknowledges forest communities as integral to the survival and sustainability of forests. The FRA provides for both individual and community rights, and statutory backing for community-driven forest governance (Kumar and Kerr, 2012).

The FRA emphasizes the authority of Gram Sabhas or village councils, which consist of all adult members of the village, for self-governance, including the right to decide about mining on their lands, among forest-dwelling communities (Chowdhury, 2016).[5] The autonomy of Gram Sabhas is understood as fundamental for democracy in forested Adivasi areas. Consequently, the FRA, which gives rights and decision-making powers to village councils, was seen as a hard-won democratic right for forest communities (Kothari, 2016).

The FRA reflected normative changes in international Indigenous rights through presupposing FPIC of communities (Sigamany, 2022). However, the FPIC clause was only subsequently appended through a 2009 circular issued by the Ministry of Environment. The circular directed state governments to show documented proof that communities' forest rights had been settled, and that Gram Sabhas had given their FPIC, before industrial projects could be approved in forests. It specified a minimum quorum of 50 per cent attendance in village council meetings for decisions to be made (Ministry of Environment and Forests, 2009). By giving decision-making powers to forest-dwelling communities, the FRA offered an alternative paradigm to centralized forest management and forest-based climate action. Its proper implementation could improve communities' living conditions (Agarwal and Dash, 2022).

But the opportunity of forest rights has largely been lost. Analysis of the FRA's implementation in the first ten years has found exceptionally low rates of recognition, particularly for community rights – a mere 3 per cent across India (CFR-LA, 2016). In their paper titled 'Manufacturing consent', based on closely observed cases, journalists Chitrangada Chowdhury and Aniket Aga write that this bureaucratic process of securing consent renders the consent provision of the FRA rife for sabotage by the state-capital nexus (2020, p 77). Bureaucracies in central Indian states are loath to give up control over forests, owing both to a persistent colonial mindset and for mining revenues, resulting in poor implementations, violation of community consent, even overturning previously granted forest rights, and the dilution of the FRA's provisions. Chowdhury and Aga conclude that the sabotage of the consent process points to a bigger conflict between self-determination for communities and eminent domain exercised by the state-corporate complex that needs to be addressed (2020, p 80).[6]

'Democracy Zindabad!' *(Long Live Democracy): Mahan's fight for forest rights*

In 2015, a grassroots movement in the Singrauli district in the central Indian state of Madhya Pradesh – a region known as the energy capital on account of producing 10 per cent of India's thermal power – saved their forests from coalmining. The four-year resistance in the Mahan forests on the fringes of one of India's largest coal basins ended when the Indian government, acting on a Supreme Court order, cancelled the coalmine. Mahan's coalmining conflict began in 2006 when the central government allocated the Mahan coalmine to a joint venture between Essar Power, one of India's largest thermal power producers, and Hindalco Ltd, an Indian aluminium manufacturing company, through a corrupt auctioning process. In 2012, around the time when the state granted forest clearance without consent in preparation for mining, a collaboration between Greenpeace India and a local movement organized under the banner of the Mahan Sangharsh Samiti (MSS) or Mahan Resistance Front came together to fight against the approval and demand forest rights under the FRA (Pillai, 2017).

Mahan Sangharsh Samiti and Greenpeace

Greenpeace's international climate campaign focused on 'keeping coal in the ground' to stop climate change. Greenpeace India had therefore been campaigning against new coalmines in central Indian forests since 2004 (Fernandes, 2012). Keeping in mind the context in central India, of mining without Adivasi consent, Greenpeace worked to raise awareness among local communities about their forest rights and authority over mining activities on their lands (Talukdar, 2019). Initial meetings organized in the villages around Mahan in 2011 helped Greenpeace understand how the community saw their ownership over forests. Ground-level surveys revealed forests to be indispensable to livelihoods, yet locals remained unaware of their forest rights. A report based on these surveys concluded that: 'The process of recognition of rights, would … allow for communities to believe in their rightful claims over forests, and presumably translate into forest dwelling communities feeling the need to hold on to what they are able to recognise as theirs' (Kohli et al, 2012, p 8).

During my ethnographic research in Mahan, I found that the mobilization of this forest constituency was motivated by a growing discontent with the state and company's interference in their everyday lives. People were especially motivated to fight the state's corrupt behaviour once they learnt they had legal rights over their local forests.

The state-corporate nexus and forging consent

The Madhya Pradesh state government was legally required to conduct *Gram Sabha* meetings across 54 potentially impacted villages around Mahan to determine whether the community consented to mining before granting forest clearance. However, the local administration and company agents allegedly disrupted *Gram Sabha* meetings to prevent people from registering their community forest rights under the FRA (Talukdar, 2017).

While talking to movement leaders, they reported that the specific incident that shaped Mahan's resolve to fight the coalmine occurred in 2013, when local government and company agents forged signatures on a referendum on coalmining. The episode became public knowledge only after Greenpeace obtained documents on the *Gram Sabha* proceedings through the right to information process (Pioneer, 2014). MSS members said they grew suspicious that evening after seeing the local administrator and police forcing people to sign the resolution after the *Gram Sabha* had concluded. They said that they suspect that signatures were forged on a massive scale later at night when company officials were seen at the village chief's residence. MSS members told me they were shocked to see their own names on the resolution document once it was obtained through a right to information request.[7]

Bribery is a common tactic deployed by mining corporations to manufacture consent (Chowdhury and Aga 2020). Acting on the directive of the federal Ministry for Tribal Affairs to conduct fresh Gram Sabhas, the district administration set the new village council dates. But MSS leaders have alleged that company agents bribed and intimidated villagers ahead of the meetings, including by distributing chicken and alcohol among the men and saris among women, visiting people's homes to compel a vote for mining and promising jobs when the coalmines open. While the state government forged signatures in Amelia, Mahan's largest village, as revealed through documents obtained by Greenpeace under right to information, it altogether avoided the community consent process in 53 other potentially affected villages around Mahan (Greenpeace India, 2014).

As reported in the national media, the Ministry for Environment and Forests issued a final forest clearance in 2014, on the basis of the fraudulent village council resolution (Press Trust of India, 2014). Taking into consideration these allegations of everyday acts of sabotage, violence, deceit and misinformation, it is fair to argue in the same vein as Chowdhury and Aga

(2020) that the state-corporate apparatus undermines the FRA's objective of redressing colonial injustices towards forest-dependent communities.

Forests and rights at all costs

Despite alleged threats and intimidations, three villages including Amelia managed to file for forest rights by 2016. Based on an understanding that their historic connection with the forest is now recognized by law, the word *adhikaar* or 'right' entered the movement lexicon. The cancellation of the coalmine in 2014 brought a sense of empowerment, and people celebrated the event as a victory of democracy on the ground. In Greenpeace India's New Delhi office, I saw a framed photograph of MSS members standing under a giant Mahua tree, taken on the day of the cancellation. They are holding a banner that says 'Loktantra Zindabad' in Hindi, meaning 'long live democracy'. The Mahua tree is an economic and cultural nerve centre for central Indian forest communities, who collect and sell its fruit and flowers, which are used for medicinal purposes.

I was present at the second anniversary celebrations of the coalmine cancellation held in Amelia village in March 2017, where men, women and children chanted slogans and sang about forest rights. To my remark that the landscape, enclosed by the Mahan River and undulating forested hills, was beautiful, one of the MSS leaders said conclusively, 'Jo bhi haye, hamara haye' (it doesn't matter whether it is beautiful, what is important is that it is ours)!

The politics of native title in Australia and forest rights in India: similarities and differences

The experiences of the W&J with native title in Australia and Mahan with forest rights in India reveal similarities and differences in terms of the intent and extent of Indigenous land rights as well as the state's role in enabling community rights versus favouring mining in a settler colonial Global North democracy and a postcolonial Global South democracy. While Australian native title has given Aboriginal Australians a voice in mining-related environmental conflicts while limiting their say on what happens on their land by excluding the FPIC, the Indian Forest Rights Act has given autonomy to village councils to self-determine on mining issues on Indigenous lands.

Yet this difference in the extent of rights enabled legally is perhaps undercut by ineffectual implementation and high levels of violation in the Indian case, due to numerous factors, including poor awareness among communities, a strong lack of political will to decentralize forest management, alongside the states' interest in facilitating mining (Kohli et al, 2012; Lee and Wolf, 2018). Therefore, while native title covers over 32 per cent of the Australian landmass after 30 years of the NTA's enactment, in India, slow and flawed implementation and high rejection rates of community forest rights claims (as high as 50 per cent in central India) has meant that only 3 per cent of the potential for community forest rights (excluding India's northeastern region) has been realized in the ten years of the FRA. Overall, only 14.5 per cent of the minimum potential forest areas for forest rights have been recognized (CFR-LA, 2016).[8] A structural divide in the state's implementation of Indigenous land rights between Australia and India persists.

This structural divide speaks to a North-South distinction that Williams and Mawdsley (2006) argue can be experienced even in the case of democracies. Marginalized communities in the Global North experience challenges in terms of accessing rights and justice, such as ongoing experiences of denial of sovereign rights among Indigenous Australians (Moreton-Robinson, 2017). But the combined realities of poverty and everyday violence from the developmental state as experienced by Adivasis can make their structural and procedural injustices incomparable with that of Northern Indigenous Peoples (Baviskar, 2012).

We need to understand a key difference between the Australian and Indian cases of Indigenous self-determination through this argument: in India, despite the law enshrining the provision for self-determination and FPIC, it is overwhelmed by legal violations that ultimately serve to deny rights and historic justice to communities. Given these different contextual challenges – in settler-colonial Australia, the limitations of native title itself, whereas in postcolonial India, extra-legal challenges relating to the implementation and exercizing of community rights – the fight for self-determination by the W&J and Mahan assumed different dimensions. While in both cases climate justice signified being able to say no to mining on their lands, the former's approach required exposing the limits of native title, while the latter required asserting democratic rights to their state-enshrined legal rights.

Administrative and legal procedures and actions of governments in both Australia and India reflect a larger process within a neoliberal economic framework that prioritizes mining over Indigenous rights, as Chowdhury and Aga (2020) argue. Both the W&J and Mahan cases demonstrate the efforts of manufacturing of Indigenous consent and a state-corporate nexus working to deny Indigenous community rights including through violations and amendments of legal frameworks.[9]

Given the structural divide discussed earlier, the ability of communities fighting mining to access justice and visibility for their struggles differ in scale and degree across these places, as do various aspects of bureaucratic, administrative and procedural injustice experienced by communities. These qualitative differences unfold within an overarching neoliberal framework, which is operative in both places and favours mining over Indigenous self-determination. I highlight two significant qualitative differences between the two cases concerning communities' awareness of rights and the relative autonomy of the movements.

The W&J's resistance is seen as an outstanding Indigenous movement for self-determination. Mahan, on the other hand, is yet another central Indian community without a history of prior mobilization or awareness of its forest rights. The multiyear process undertaken by Greenpeace to raise community awareness on forest rights demonstrates a fundamental challenge for grassroots and Indigenous mobilizations for self-determination in central India. Consequently, it raises the question of their critical need for support from mainstream NGO and environmental groups.

The extent of autonomy in the W&J's movement, its strategic collaboration with the environmental movement, and its ability to access the UN to appeal for Indigenous self-determination and expose Australia's violations stand in contrast with the co-dependency with which the Mahan mobilization was built up through Greenpeace's support. This co-dependency was evident at other stages too, not just at the initial stages of fostering forests rights awareness: Greenpeace supported the Mahan community with freedom of information requests (among many other tools and strategies) to discover the forgeries of *Gram Sabha* resolutions. The extent of this co-dependent relationship signifies challenges for India's Indigenous communities in accessing platforms for appeal and justice on a par with their Global North counterparts. It further qualifies my last point about the needs for grassroots struggles for critical support from mainstream, urban activists to access justice.

To conclude this section, the two cases demonstrate how FPIC is being mobilized by Indigenous struggles against coalmining in the Global North and South, a settler colonial and a postcolonial context, in two democracies. The W&J and Mahan cases show similarities primarily in the tension between the intent of land rights and self-determination and state favouritism of mining, underscoring the need for solidarity for these crucial human rights struggles in the global fight for climate justice. But the two cases reveal structural differences that provide insights for how solidarity needs to be considered differently in the two contexts. These differences across Australia and India, under the overarching similarity of neoliberal forces, help to qualify how solidarity can be considered in a Global North versus a Global South context, which I will discuss next.

Recommendations: solidarity and North–South intersectionality

The present aim of global climate activism of 'keeping coal in the ground' is entangled with those of Indigenous communities at the frontline of fossil fuel extraction, with whom they are forming alliances. Extraction from Indigenous lands occurs not just through mining (for fossil fuels and minerals), but also through other modes of development without Indigenous consent, including 'clean' and renewable energy. Activists can therefore consider broadening the objectives and temporal scope of building solidarity with frontline Indigenous land struggles.

Looking at alliances between environmentalists and Indigenous groups in Australia in an earlier period from today's climate activism (roughly from 1990 to 2010), Vincent and Neale (2016) describe them as 'tactical alliances': proximal but unstable relations built around a short-term, shared environmental outcome, in most cases to stop mining. They argue that owing to a fundamental mismatch of visions, deeper collaborations have not been possible.

The W&J entered into a 'tactical alliance' with the mainstream Stop Adani movement and asserted the distinctiveness of their campaign's historic significance and their relative autonomy. But Mahan showed a stronger co-dependence with Greenpeace India. Greenpeace's campaign goals had a tactical focus on stopping coalmining in alliance with frontline communities. Yet the social and political context in central India, where communities had to be first made aware of their legal rights, necessitated this international environmental NGO to take a longer, layered approach in India. The two cases point to specific ways in two different contexts for solidarity making by global climate activism with First Nations struggles.

The scale and structure of the W&J's and Mahan's challenges are different. Mahan faced the prospect of a direct loss of livelihoods from losing forest and lands on which they depend for economic survival. But in both cases, communities fought to secure critical human rights – to survival (including cultural survival), livelihood, healthy environment and freedom from fear of ecological destruction. These critical human rights are becoming central to demands for global climate justice (Skillinton, 2017). Increasingly, Indigenous ownership and governance of forests and land is being seen as crucial for climate justice (for communities) and better climate outcomes (for the planet) (Agarwal and Dash, 2022). Finding common cause with these human rights questions can help to reorient global climate activism towards building deeper and intersectional solidarities with the communities with they are engaged.

I propose two substantive recommendations for practitioners and scholars around centring Indigenous land rights and self-determination – effectively their human rights – in a global outlook of climate activism. First, take past, present and futures into account. The new approach of environmentalism, predicated on solidarity with frontline communities, can focus on systemic and structural causes for human rights violations in the political economy of energy systems. It can broaden and deepen the grounds of its advocacy to ensure that the postcarbon and renewables-powered society it envisions do not replicate the systemic human rights concerns of fossil-fuel regimes. During my fieldwork in Mahan, I learnt that many of the families had already been twice displaced by large dams built in the 1980s in the region, settling in Mahan where the risk of coalmining induced displacement visited them in 2012. Sustained solidarities with Indigenous self-determination struggles that foreground just and secure Indigenous futures as goals can attempt to address cyclical and systemic issues that visit communities.

Second, acknowledge North-South differences. A global understanding of climate justice needs to acknowledge critical differences experienced by Indigenous struggles across the Global North and South that William and Mawdsley (2006) refer to as a structural divide. Climate justice scholars need to avoid transplanting Northern climate justice discourse to the South, which can hamper the building of a clear understanding about the relation between poverty and environmental justice (Lawhon, 2013). Like many subsistence communities, Mahan residents mostly still live without electricity. Subsistence based communities in the Global South are doubly disadvantaged through a lack of energy access and environmental injustice. Instead of decolonized movements for sovereign self-determination in the settler colonial Global North, as seen in the case of the W&J, their challenges need to be considered within a postcolonial development paradigm: of communities asserting their democratic rights as citizens through fighting for forest rights. Struggles for claiming forest rights and asserting democracy on the ground are means for claiming a stake in development, which bypasses them while dispossessing them of their *jal*, *jangal* and *jameen*.

Notes

[1] The NTA recognized that native title, essentially a form of Indigenous land title, may continue to exist in areas where Indigenous people still occupied and could display a continuing association with their precolonial traditional land, with the caveat that the rights of native title holders would have to yield in the event of conflicts with non-Indigenous interests.

[2] Independent analysis showed that the Adani deal was one of the worst for Traditional Owners (Meaton, 2017), with compensations less than half the industry average (Quiggin, 2017; Robertson, 2017).

[3] The 2017 *McGlade decision* delivered by the Western Australian Federal Court determined that the signed consent of all members of a native title applicant group is required for an ILUA to be valid for registration under the NTA (McHugh, 2017). The *McGlade decision* could invalidate Adani's contested ILUA with the W&J, but the Federal Government proceeded to immediately amend the Native Title Act to overturn the *McGlade decision*.

[4] India's Indigenous people are designated as Scheduled Tribes. The term 'Adivasi' means 'native dweller' and the term is used by Indigenous tribes in central India. The term was coined in the 1930s through political movements for Indigenous identity in central India.

[5] Prior to the FRA, the Panchayat (Extension to Schedule Areas) Act in 1996 (PESA) had extended powers of self-governance to Gram Sabhas or village councils in Scheduled Areas.

[6] The term 'eminent domain' refers to the overarching power vested in the state to compulsorily acquire private property or land without consent for public use.

7   The state government was asked to conduct fresh council meetings to determine people's consent after Greenpeace and MSS took the matter to the Ministry for Tribal Affairs in the central government. The central government ministry expressed concerns at violations to the FRA both in Mahan and more widely in central India (Sethi, 2015).
8   A 2021 circular by the Indian government to state governments gives the latter the responsibility to review and facilitate the proper implementation of forest rights, stating that 'despite a considerable lapse of time since it (FRA) came into force, the process of recognition of forest rights is yet to be completed' (Nandi, 2021).
9   Compromises to the provisions of the FRA and rights of communities have also arisen from the shortening of approval timelines for environmental clearances, and relaxing the requirement of public consent for coal mining projects under the FRA (CSE, 2016). To boost coalmining, India loosened forest clearance processes in 2020 that compromised the autonomy of Gram Sabhas (CFR-LA, 2020). The democratic provisions and intent of the FRA to redress historic injustices have been further diluted by the Indian Narendra Modi government in 2022 through the notified new Forest Conservation Rules, allowing people's forests to be diverted for projects even without a No Objection Certificate from *Gram Sabhas*.

References

Agarwal, S. and Dash, T. (2022) *Policy Brief: Securing Climate Justice for India's Forest-Dependent Communities*, Available from: http://fra.org.in/document/Revised%20Policy%20brief%20-%20Securing%20climate%20justice%20for%20forest-dependent%20communities.pdf [Accessed 20 November 2022].

Altman, J. (2009) 'Indigenous communities, miners and the state in Australia', in J. Altman and D. Martin (eds) *Power, Culture and Economy*, Canberra: Australian National University Press, pp 17–50.

Altman, J. (2012) 'Indigenous rights, mining corporations, and the Australian state in the politics of resource extraction 2012', in S. Sawyer and E. Gomez (eds) *The Politics of Resource Extraction; Indigenous Peoples, Multinational Corporations, and the State*, London: Palgrave Macmillan, pp 46–74.

Australian Associated Press (2016) 'Adani's Carmichael coalmine leases approved by Queensland', 3 April, Available from: https://www.theguardian.com/environment/2016/apr/03/adanis-carmichael-coalmine-leases-approved-by-queensland [Accessed 20 March 2020].

Australian Institute of Aboriginal and Torres Strait Islander Studies (AIATSIS) (2022) 'The Native Title Act', Available from: https://aiatsis.gov.au/about-native-title [Accessed 20 September 2022].

Baviskar, A. (2012) 'Extraordinary violence and everyday welfare: the state and development in rural and urban India', in S. Venkatesan and T. Yarrow (eds) *Differentiating Development: Beyond an Anthropology of Critique*, Oxford: Berghahn Books, pp 126–144.

Bebbington, A. et al (2008) 'Contention and ambiguity: mining and possibilities of development', *Development and Change*, 39(6): 887–914.

Birch, T. (2018) '"On what terms can we speak?" Refusal, resurgence and climate justice', *Coolabah*, 24-25: 2–16.

Brigg, M. (2018) 'Killing Country (part 5): Native Title colonialism, racism and mining for manufactured consent', *New Matilda*, 30 January, Available from: https://newmatilda.com/2018/01/30/native-title-colonialism-racism-adani-and-the-manufacture-of-consent-for-mining/ [Accessed 20 September 2020].

Burragubba, A. (2016) 'Adrian Burragubba: the struggle to save country from Adani', *Green Left Weekly*, 15 April, Available from: https://www.greenleft.org.au/content/adrian-burragubba-struggle-save-country-adani/ [Accessed 20 September 2020].

Carey, M. (2019) 'An inside look at how Adani dealt with Traditional Owners', *NITV News* [online], 16 May, Available from: https://www.sbs.com.au/nitv/nitv-news/article/2019/05/16/inside-look-how-adani-dealt-traditional-owners [Accessed 20 September 2020].

Carino, J. (2005) 'Indigenous peoples' right to free, prior and informed consent: reflections on concepts and practice', *Arizona Journal of International and Comparative Law*, 22(1): 19–39.

Centre for Science and Environment (CSE) (2016) 'Report card: environmental governance under NDA government', *Down to Earth* [online], 22 June, Available from: https://www.downtoearth.org.in/coverage/governance/report-card-environmental-governance-under-nda-government-54359 [Accessed 20 September 2019].

Chowdhury, C. (2016) 'Making a hollow in the forest rights act', *The Hindu* [online], 7 April, Available from: https://www.thehindu.com/opinion/columns/Making-a-hollow-in-the-Forest-Rights-Act/article14226592.ece [Accessed 20 September 2019].

Chowdhury, C. and Aga. A. (2020) 'Manufacturing consent: mining, bureaucratic sabotage and the Forest Rights Act in India', *Capitalism, Nature, Socialism*, 31(2): 70–90.

Community Forest Rights-Learning and Advocacy (CFR-LA) (2016) *Promise and Performance, Ten Years of the Forest Rights Act in India*, Citizen's Report on the Promise and Performance of Scheduled Tribes and Other Traditional Forest Dwellers (Recognition of Forest Rights) Act 2006, Available from: https://rightsandresources.org/publication/promise-performance-forest-rights-act-2006-tenth-anniversary-report/ [Accessed 12 March 2023].

Community Forest Rights-Learning and Advocacy (CFR-LA) (2020) *Community Forest Rights and the Pandemic: Gram Sabhas Lead the Way*, October, Available from: https://rightsandresources.org/wp-content/uploads/2020/10/CFR-and-the-Pandemic_GS-Lead-the-Way-Vol.2_Oct.2020.pdf/ [Accessed 12 March 2023].

Coyne, B. (2017) 'Re-greening rights indigeneity, climate change, and a timely re-confluence of human rights and the environment', keynote address, 2017 Environment and Planning Law Association (NSW) Annual Conference.

Crenshaw, K.W. (2017) *On Intersectionality: Essential Writings*, New York: New Press.

Food and Agricultural Organization (FAO) (2017) '6 ways Indigenous peoples are helping the world achieve #ZeroHunger', Available from: https://www.fao.org/zhc/detail-events/en/c/1028010/ [Accessed 20 November 2022].

Fernandes, A. (2012) 'How coal mining is trashing tigerland', *Greenpeace India*, 1 August, Available from: https://www.greenpeace.org/india/en/publication/984/how-coal-mining-is-trashing-tigerland/ [Accessed 14 August 2019].

Frost, M. (2016) *End of Mission Statement by Michael Frost, United Nations Special Rapporteur on the situation of human rights defenders* [visit to Australia], 18 October, Available from: https://www.ohchr.org/en/statements/2016/10/end-mission-statement-michel-forst-united-nations-special-rapporteur-situation [Accessed 20 March 2020].

Giacomini, G. (2022) *Indigenous Peoples and Climate Justice: A Critical Analysis of International Human Rights Law and Governance*, Cham: Palgrave Macmillan.

Gilbert, K. (1994) *Because a White Man'll Never Do It*, Pymble, NSW: Angus & Robertson.

Greenpeace India (2014) 'Civil society urges the centre not to trample over the rights of forest dwellers in Mahan. Implement FRA in 54 villages of Mahan, demand Mahan Sangharsh Samiti', Press Release, 1 August, Available from: https://www.greenpeace.org/india/en/press/2537/civil-society-urges-the-centre-not-to-trample-over-the-rights-of-forest-dwellers-in-mahan-implement-fra-in-54-villages-of-mahan-demand-mahan-sangharsh-samiti/ [Accessed 20 September 2019].

Klein, N. (2016) 'Let them drown: the violence of othering in a warming world', *London Review of Books*, 2 June.

Kohli, K., Kothari, A. and Pillai, P. (2012) 'Countering coal?' Discussion paper, Kalpavriksh and Greenpeace India, Available from: https://www.greenpeace.org/india/en/publication/989/countering-coal-community-forest-rights-and-coal-mining-regions-of-india/ [Accessed 15 September 2019].

Kothari, A. (2016) 'Decisions of the people, by the people, for the people', *The Hindu* [online], 18 May, Available from: https://www.thehindu.com/opinion/op-ed/Decisions-of-the-people-by-the-people-for-the-people/article14324692.ece [Accessed 20 September 2019].

Kumar, K. and Kerr, J.M. (2012) 'Democratic assertions: the making of India's recognition of Forest Rights Act', *Development and Change*, 43(3): 751–771.

Lahiri-Dutt, K. (2016) 'The diverse worlds of coal in India: energising the nation, energising livelihoods', *Energy Policy*, 99: 203–213.

Langton, M. (2012) *Boyer Lectures 2012: The Quiet Revolution: Indigenous People and the Resources Boom*, Sydney: ABC Books.

Lawhon, M. (2013) 'Situated, network environmentalism: a case for environmental theory from the South', *Geography Compass*, 7(2): 128–138.

Lee, J. and Wolf, S.A. (2018) 'Critical assessment of implementation of the Forest Rights Act of India', *Land Use Policy*, 79: 834–844.

Lyons, K. (2017a) 'The Queensland government is the real driver in Adani's dirty land grab', *New Matilda*, 22 November, Available from: https://newmatilda.com/2017/11/22/the-queensland-government-is-the-real-driver-in-adanis-dirty-land-grab/ [Accessed 20 March 2020].

Lyons, K. (2017b) 'Traditional owners expose Adani's relentless pursuit of W&J country', *New Matilda*, 23 November, Available from: https://newmatilda.com/2017/11/23/traditional-owners-expose-adanis-relentless-pursuit-of-wj-country/ [Accessed 15 March 2020].

Lyons, K. (2019) 'Securing territory for mining when Traditional Owners say "no": the exceptional case of Wangan and Jagalingou in Australia', *The Extractive Industries and Society*, 6: 756–766.

Lyons, K., Esposito, A. and Johnson M. (2021) 'Intervention – the pangolin and the coal mine: challenging the forces of extractivism, human rights abuse, and planetary calamity', *Antipode Online*, 1 February, Available from: https://antipodeonline.org/2021/02/01/the-pangolin-and-the-coal-mine/ [Accessed 12 March 2024].

Mabo and Others vs The State of Queensland [No. 2] 1992, Available from: https://aiatsis.gov.au/explore/mabo-case [Accessed 29 March 2024].

Market Forces (2015) 'Funding the Galilee coal mines just got harder', *Market Forces*, 15 June, Available from: https://www.marketforces.org.au/funding-the-galilee-coal-mines-just-got-harder/ [Accessed 20 March 2020].

McHugh, B. (2017) 'Turmoil over Indigenous land use ruling', *Australian Broadcasting Corporation Rural*, 8 February, Available from: https://www.abc.net.au/news/rural/2017-02-08/turmoil-over-indigenous-land-use-ruling/8250952 [Accessed 20 March 2020].

Meaton, M. (2017) 'Adani Carmichael coalmine ILUA Assessment', *Economic Consulting Services*, Available from: http://wanganjagalingou.com.au/wp-content/uploads/2017/12/Adani-Coal-presentation-2December2017.pdf [Accessed 20 March 2020].

Ministry of Environment and Forests (2009). 'Diversion of forest land for non-forest purposes under the Forest (Conservation) Act, 1980 – ensuring compliance of the Scheduled Tribes and Other Traditional Forest Dwellers (Recognition of Forest Rights) Act 2006', *Circular*, 7 July, Available from: https://forestsclearance.nic.in/writereaddata/public_display/schemes/981969732$3rdAugust2009.pdf [Accessed 22 November 2022]

Moreton-Robinson, A. (2017) 'Citizenship, exclusion and the denial of Indigenous sovereign rights', *Australian Broadcasting Corporation Religion and Ethics*, 30 May, Available from: https://www.abc.net.au/religion/citizenship-exclusion-and-the-denial-of-indigenous-sovereign-rig/10095738 [Accessed 10 June 2021].

Mookerjea, S. (2019) 'Renewable energy transition under multiple colonialisms: passive revolution, fascism redux and utopian praxes', *Journal of Cultural Studies*, 33(3): 570–593.

Nandi, J. (2021) 'Review implementation of forest rights: union ministries to state governments', *Hindustan Times*, 7 July, Available from: https://www.hindustantimes.com/environment/review-implementation-of-forest-rights-union-ministries-to-state-govts-101625562374847.html [Accessed 20 November 2022].

Norman, H. (2015) *What Do We Want? A Political History of Aboriginal Land Rights in New South Wales*, Canberra: Aboriginal Studies Press.

Norman, H. (2016) 'Coal mining and coal seam gas on Gomeroi country: sacred lands, economic futures and shifting alliances', *Energy Policy*, 99: 242–251.

Perera, J. (2016) 'Old wine in new bottles: self-determination, participatory democracy and free, prior, and informed consent', in P. Sillitoe (ed) *Indigenous Studies and Engaged Anthropology*, Abingdon: Routledge, pp 147–162.

Pillai, P. (2017) 'An ongoing battle', *Frontline*, 15 September, Available from: https://frontline.thehindu.com/social-issues/an-ongoing-battle/article9831547.ece [Accessed 27 September 2019].

Pioneer (2014) 'HC asks SP to probe Gram Sabha forgery', *Daily Pioneer*, 16 July.

Porter, L., Bosomworth, K., Moloney, S. and Naarm, B. (2020) 'Decolonising climate change adaptation', *Planning Theory and Practice*, 21(2): 293–321.

Press Trust of India (2014) 'Moily planning to give oil field to Essar cheaply, alleges AAP', *Economic Times* [online], 17 April, Available from: https://economictimes.indiatimes.com/news/politics-and-nation/moily-planning-to-give-oil-field-to-essar-cheaply-alleges-aap/articleshow/33828233.cms [Accessed 27 September 2019].

Quiggin, J. (2017) 'The numbers don't stack up: W&J's rights on the chopping block for Adani's non-viable project', *New Matilda*, 24 December, Available from: https://newmatilda.com/2017/12/24/the-numbers-dont-stack-up-wjs-rights-on-the-chopping-block-for-adanis-non-viable-project/ [Accessed 15 March 2020].

Ramesh, J. and Khan, M.A. (2015) *Legislating for Justice: The Making of the 2013 Land Acquisition Law*, New Delhi: Oxford University Press.

Rights and Resources Initiative (2017) *Annual Narrative Report*, Available from: https://rightsandresources.org/wp-content/uploads/2018/04/2017 _RRI_Annual-Narrative-Report.pdf [Accessed 20 July 2022].

Robertson, J. (2017) 'Adani's compensation well below industry standard, report finds', *ABCNews* [online], 1 December, Available from: https:// www.abc.net.au/news/2017-12-01/adani-compensation-well-below-industry-standard-report-finds/9212058 [Accessed 20 June 2020].

Robertson, J. and Sigato, T. (2018) 'Adani Indigenous challenge dismissed by Federal Court, Government could cancel mine Native Title', *ABC News* [online], 17 July, available from: https://www.abc.net.au/news/2018-08-17/adani-federal-court-traditional-owners-nativ e-title/10131920 [Accessed 20 September 2020].

SBS News (2017) 'Labor slams government over Native Title move', 16 February, Available from: http://www.sbs.com.au/news/article/2017/02/ 16/labor-slams-govt-move-native-title [Accessed 20 March 2020].

Sethi, N. (2015) 'NDA govt's grouse with Greenpeace: Mahan coal block protests', *Business Standard* [online], 20 February, Available from: https:// www.businessstandard.com/article/economy-policy/nda-govt-s-gro use-with-greenpeace-mahan-coal-block-protests-115022000023_1.html [Accessed 20 September 2019].

Short, D. (2007) 'The social construction of Indigenous "Native Title Land Rights" in Australia', *Current Sociology*, 55(1): 857.

Shrivastava, A. and Kothari, A. (2012) *Churning the Earth: The Making of Global India*, New Delhi: Penguin.

Sigamany, I. (2022) 'The Indian Forest Rights Act (2006) and international legal regimes: a review and comparison', in Z. Mohammad, N. Reshmy and G. Shi (eds) *Resettlement in Asian Countries: Legislation, Administration, and Struggles for Forest Rights*, New York: Routledge, pp 244–256.

Skillinton, T. (2017) *Climate Justice and Human Rights*, London: Palgrave Macmillan.

Stephenson, W. (2015) *What We're Fighting for Now Is Each Other: Dispatches from the Front Lines of Climate Justice*, Boston, MA: Beacon Press.

Talukdar, R. (2017) 'Hiding neoliberal coal behind the Indian poor', *Journal of Australian Political Economy*, 78: 132–158.

Talukdar, R. (2019) 'Profit before people: why India has silenced Greenpeace', *New Matilda*, 29 March, Available from: https://newmatilda. com/2019/03/29/profit-before-people-why-india-has-silenced-greenpe ace/?fbclid=IwAR2NwczXun5q1Ug-g3kSxvaOJLInTfwNlnS6Up3X xts7A7XZF_wbvOsHKYc [Accessed 20 June 2020].

Talukdar, R. (2021) 'Cutting carbon from the ground up: an ethnography of anti-coal movements in India and Australia', PhD Thesis, Faculty of Arts and Social Sciences, University of Technology Sydney.

Turner, B.S. and Rojeck, C. (2001) *Society and Culture: Principles of Scarcity and Solidarity*, London: Sage.

United Nations Declaration of the Rights of Indigenous Peoples (2007) Available from: https://social.desa.un.org/issues/indigenous-peoples/united-nations-declaration-on-the-rights-of-indigenous-peoples [Accessed 20 September 2022].

United Nations Department of Economic and Social Affairs (2021) *State of the World's Indigenous Peoples: Rights to Land, Territories and Resources*, New York, Volume 5.

Vincent, E. and Neale, T. (2016) *Unstable Relations: Indigenous People and Environmentalism in Contemporary Australia*, Crawley, WA: UWA Publishing.

Wagner, M. (2017) *Australia's Ongoing Violation of the Rights of the Wangan and Jagalingou People to Be Consulted in Good Faith about the Development of the Native Title Amendment (Indigenous Land Use Agreements) Bill 2017 (Cth) and Its Impacts on the Wangan and Jagalingou, Earth Justice*, 1 March, Available from: http://wanganjagalingou.com.au/wp-content/uploads/2018/08/Request-for-Urgent-Action-by-Wangan-and-Jagalingou-People-to-CERD-31-July-2018.pdf [Accessed 20 September 2020].

W&J (Wangan & Jagalingou) (2015) *Submission to the Special Rapporteur on Indigenous Peoples by the Wangan and Jagalingou People*, 2 October, Available from: http://wanganjagalingou.com.au/wp-content/uploads/2015/10/Submission-to-the-Special-Rapporteur-on-Indigenous-Peoples-by-the-Wangan-and-Jagalingou-People-2-Oct-2015.pdf [Accessed 20 September 2020].

W&J (2016) '"We stand in the way" of Adani mine say Traditional Owners', *Wangan and Jagalingou Family Council*, 5 December, viewed 20 September 2020, Available from: https://wanganjagalingou.com.au/we-stand-in-the-way-of-adani-mine-say-traditional-owners-seek-urgent-meeting-with-gautam-adani-after-filing-objection-to-carmichael-mine-land-use-agreement/ [Accessed 20 September 2020].

W&J (2018) 'The path of resistance', *Wangan and Jagalingou Family Council*, 11 July, Available from: http://wanganjagalingou.com.au/the-path-of-resistance/ [Accessed 20 March 2020].

Whyte, K. (2021) 'Time as kinship', in J. Cohen and S. Foote (eds) *The Cambridge Companion to Environmental Humanities*, Cambridge: Cambridge University Press, pp 39–55.

Williams, G. and Mawdsley, E. (2006) 'Postcolonial environmental justice: government and governance in India', *Geoforum*, 37: 660–670.

7

Popular Intellectuals, Social Movement Frames and the Evolution of the Anti-Mining Movement in the Niyamgiri Mountains, Odisha, India

Souvik Lal Chakraborty and Julian S. Yates

Introduction

The eastern state of Odisha in India has been facing the problem of development-induced human displacement since the early years of independence in 1947 (Pattnaik, 2013). This displacement is an outcome of Odisha's trajectory of economic modernization, resource-based development and investments that centre the exploitation of large deposits of strategically important minerals (Kale, 2013). The consequences of this strategy are increasingly playing out as conflict over mining and natural resource extraction (see Padel and Das, 2007, 2010; Das and Padel, 2013; 2020; Pattnaik, 2013; Padhi and Sadangi, 2020). This chapter focuses on a high-profile case of mining-related conflict: the movement to save the Niyamgiri Mountains, which are sacred to the Dongria Kondh[1] and Kutia Kondh tribal people, from mining by Vedanta Resources Ltd. Understanding the evolving trajectory of this movement has wider implications for understanding how injustice is addressed through frames of political activism in the Global South.

While some aspects of the Niyamgiri Movement are well studied – such as its international reach and attention (see, for example, Padhi and Sadangi, 2020; Borde and Bluemling, 2021) – this chapter sheds new light on the network of on-the-ground activists who helped to develop and adapt the 'collective action frames' (Snow, 2004) that propelled the movement. As these frames struggle to account for diverse and alternative views of development, questions remain as to the ongoing efficacy of the movement, despite achieving a favourable judgment from the Supreme Court of India in 2013, which has prevented mining to this day. This chapter explains how social movement frames are central to understanding some of the movement's key successes and its relative stagnation in recent years.

The Niyamgiri Movement is contextualized by deep inequality and poverty, despite the great resource wealth of Odisha. Odisha has a deposit of 51 per cent of India's bauxite, 96 per cent chromite, 92 per cent nickel, 33 per cent iron ore, 24 per cent coal and 43 per cent manganese ore (Government of Odisha, 2021), but dependency on the mining sector for growth and development has come with social and ecological costs (Sahu, 2019). The minerals are located in areas mostly inhabited by people of the tribal communities, and this has led to conflict with mining companies and government institutions that support mining-led development (Oskarsson, 2013; Sahu, 2019; Das and Padel, 2020). Similar to documented case studies of extractivism around the globe (see Perreault, 2013; Sahu, 2019; Das and Padel, 2020), the lives and livelihoods of people living in remote areas of Odisha are impacted by mining, and in most cases people do not receive proper or adequate compensation (Lahiri-Dutt, 2014).

A bauxite mining project proposed by mining company Vedanta Resources Ltd cuts through the deep gorges and cascading streams of the densely forested Niyamgiri Mountains. The location of the proposed mining site poses a threat to the lives and livelihoods of Dongria Kondhs, Kutia Kondhs[2] and people from other vulnerable communities. The Dongria Kondhs consider Niyam Raja (The King Who Upholds the Law) as their first King and he is their spiritual sovereign (Jena et al, 2002, p 191). They consider the entire Niyamgiri Mountain ranges – spread over 250 square kilometres in the Raygada and Kalahandi districts of Odisha – as the kingdom of Niyam Raja (Jena et al, 2002; Tatpati et al, 2016). Initially the Dongria Kondhs were not aware of Vedanta's plan to mine the mountains because they were kept in the dark about the public hearing processes (Marshall and Balaton-Chrimes, 2016).

The local and Odisha based leaders who opposed the project began by establishing contacts with the Dongria Kondhs, who finally also joined the Niyamgiri Suraksha Samiti (Organization to Save Niyamgiri, hereinafter NSS) (Marshall and Balaton-Chrimes, 2016; Chakraborty, 2022). The NSS is the main organization which drives the Niyamgiri Movement. The Niyamgiri Movement gained momentum from 2004 when external activist groups supported the NSS, and a number of petitions were filed by the activists in the Supreme Court of India, acting in solidarity along with the people's resistance on the ground (Marshall and Balaton-Chrimes, 2016; Kapoor, 2017).

What began as localized place-based resistance against the Vedanta project eventually turned into a multinational resistance movement against mining and resource extraction, and it has continued as the Niyamgiri Movement over the past two decades. This chapter explores the evolving nature of the Movement, paying attention to how collective action frames react to shifting political economic contexts and the movement needs. The Niyamgiri Movement case study follows a pattern in land conflict, where 60 per cent of mining-related conflicts in India occur in areas that have a majority tribal population (Worsdell and Sambhav, 2020). Telling this story helps practitioners and movement activists understand the ways in which movements evolve as their participants and contexts change.

Frames and popular intellectuals

Several studies of the Niyamgiri Movement explicitly or implicitly refer to the importance of framing in the resistance movement (see Kumar, 2014; Krishnan and Naga, 2017; Macdonald et al, 2017; Borde and Rasch, 2018). Snow explains that collective action frames are like 'picture frames, [that] focus attention by punctuating or specifying what in our sensual field is relevant and what is irrelevant, what is "in frame" and what is "out of frame", in relation to object of orientation' (2004, p 384). These collective action frames help to conceive how social movement actors engage in the processes of framing, and what kind of resonance framing creates to affect the form and efficacy of a social movement. In the context of social movements, the analysis of collective action frames helps in understanding the evolution of a social movement. Framing activities generate 'action-oriented beliefs and meanings that legitimize and inspire social movement campaign and activities' (Snow et al, 2019, p 395).

Snow and Benford (1988, pp 199–202) identified three core framing tasks: 'diagnostic framing', 'prognostic framing' and 'motivational framing'. These three core framing tasks need to be performed to move 'people from balcony to the barricades' (Benford and Snow, 2000, p 615) – that is, they galvanize action. Diagnostic framing involves problem identification and blame attribution (Snow and Benford, 1988; Benford and Snow, 2000). The second core framing task – prognostic framing – involves 'the articulation of a proposed solution to the problem, or at least a plan of attack, and the strategies for carrying out the plan' (Benford and Snow, 2000, p 616). Motivational framing is the final core framing task which provides a 'rationale for engaging in ameliorative collective action, including the construction of appropriate vocabularies of motive' (Benford and Snow, 2000, p 617).

While framing analysis sheds light on the movement processes and dynamics, none of the existing theorizing specifically or systematically analyses the strategic and intentional processes of framing (including the evolution of frames across a social movement cycle), or the impacts of these frames on the form, function and efficacy of the movement. This process of framing, Baud and Rutten (2005) argue, is essential to delineate for building an understanding of social movements, because collective action frames help in representing the core of a movement's ideology, identity and rationale for collective action. In this chapter, we will address this gap in the existing body of academic literature by analysing the role of 'popular intellectuals' (as conceptualized by Baud and Rutten [2005]) in shaping 'collective action framing' (Snow et al, 1986; Benford and Snow, 2000) of the Niyamgiri Movement. Popular intellectuals are those individuals who may or may not be formally educated, but they try to understand the society that they want to change (Baud and Rutten, 2005). They are concerned with the plight of subaltern groups and they try to 'articulate their grievances and frame their social and political demands' (Baud and Rutten, 2005, p 2).

Baud and Rutten's (2005, p 7) emphasis is only on people who are 'framing specialists' – that is, those who 'develop, borrow, adapt and rework interpretive frames that promote collective action and that define collective interests and identities, rights and claims' (Baud and Rutten, 2005, p 7). Further, they 'tentatively' identify these individuals as 'popular intellectuals' and split them into three types: innovators, movement intellectuals and allies. Innovators are responsible for producing 'new interpretive frames and new languages for articulating collective interests, identities and claims' (Rutten and Baud, 2005, p 198). Their ideas motivate and initiate social movements, but they may remain aloof from the social movement organizations and enjoy a certain degree of autonomy (Rutten and Baud, 2005).

Drawing from the work of Eyerman (1994), Rutten and Baud (2005) have defined the second category of popular intellectuals as 'movement intellectuals' who are directly engaged in developing the collective action frame for the social movement. Eyerman (1994, p 15) defines movement intellectuals as those 'individuals who gain the status and the self-perception of being "intellectuals" in the context of their participation in political movements rather than through the institutions of the established culture'. Rutten and Baud (2005, p 199) note that the classification of movement intellectuals is a 'mixed lot'. It may include individuals who are established as intellectuals, and also individuals with little formal education. The third category of popular intellectuals are 'movement allies', which includes 'intellectuals who put their expertise and networks at the disposal of the various movements, often when these movements have already developed and received public attention' (Rutten and Baud, 2005, p 200).

In this chapter, we demonstrate the importance of the role of popular intellectuals in shaping collective action frames and analyse how the framing of the Niyamgiri Movement has evolved in the 11 years since the historical verdict of the Supreme Court of India in 2013 – the temporal point at which the Niyamgiri Movement was hailed a success internationally, but from which time activities on the ground have also continued to evolve. Many studies have explored the period of the movement up until 2013, but few studies have explored the period of the movement after 2013. We therefore contribute new insights into Niyamgiri Movement post-2013 as a case study to theorize the role of 'outsiders' in place-based movements.

Nine months in Odisha

Multisited ethnographic fieldwork for this research was conducted by the first author of this chapter (Souvik Lal Chakraborty) in India's eastern state of Odisha, from April to December 2019 as part of his PhD. Chakraborty was based in Bhubaneswar (the state capital of Odisha) and from there he travelled to several other locations to conduct interviews, including Raygada and Lanjigarh. Being the state capital, various nongovernmental organizations (NGOs) and activist organizations conduct important meetings in Bhubaneswar, providing Chakraborty with opportunities to interact with the popular intellectuals, members of the media, NGO representatives and activists focused on preventing bauxite mining in the Niyamgiri Mountains. Ethnographic fieldwork was instrumental in gaining knowledge about the nature of multiscalar activism of the popular intellectuals, but also for understanding the broader scenario of social activism in Odisha.

The heavy police and paramilitary presence in the area meant that it was not an easy task to reach the Niyamgiri Mountains. Sunil, a Bhubaneswar-based popular intellectual, played an instrumental role in arranging a visit to Niyamgiri, and even then, it was only possible to reach a few of the villages in Niyamgiri for a very short period of time. Chakraborty was nonetheless able to speak with the members of the Dongria Kondh community who are also actively involved with the NSS in Puri and in Bhubaneswar, Odisha. Sunil[3] made clear that an overnight stay in the Niyamgiri Mountains may put the community at risk, and members of the Dongria Kondh community could face questions from paramilitary forces. Therefore, to avoid putting the community at risk, Chakraborty left the village with a local contact introduced by Sunil after conducting the interviews. With deep concerns in mind, only a few interviews were conducted in villages in the Niyamgiri Mountains, and instead many interviews were conducted in Bhawanipatna (the district capital of Kalahandi), Lanjigarh (where Vedanta's aluminium refinery is located), Muniguda (Raygada district), Bhubaneswar (the state capital of Odisha) and New Delhi. Due to heavy militarization in the

Niyamgiri Mountains, Chakraborty, a relatively privileged male Indian researcher from Australia, was unable to stay in the villages in Niyamgiri, which is one of the limitations of this research. The lack of access to local communities, for example, meant that women's lived experience and perception of development in the Niyamgiri Movement are not captured in the data collection.

Twenty-five semi-structured in-person interviews aimed to identify the key actors, the relationships among various actors and their role in the Niyamgiri Movement, and the internal dynamics of the Niyamgiri Movement. The participants for the interview were chosen using a snowball sampling technique, whereby key informants assisted Chakraborty in establishing contact with additional interviewees. Five interviews were with key informants who played a significant role in the history of the Niyamgiri Movement and knew a great deal about how the movement was shaped from the day of its inception.[4]

Being in the field in Odisha for nine months provided ample time for participant observation (DeWalt and DeWalt, 2011). A field diary was maintained daily where all field observations were daily recorded. These observations helped Chakraborty to reflect, recall and more deeply understand the complex dynamics of people's movement networks in Odisha, when it came to analysing of the interviews. The identity of the participants in this research is confidential to ensure their safety and security. The fieldwork data presented in the remainder of this chapter has been mostly paraphrased and pseudonyms have been used so that the interviewees cannot be identified.

The genesis of the Niyamgiri Movement

In 1997, Sterlite Industries owned by Indian businessman Anil Agarwal signed a memorandum of understanding (MoU) with the government of Odisha to construct a mining and a refinery project in the Kalahandi district of Odisha, where the Niyamgiri Mountains are located (Marshall and Balaton-Chrimes, 2016). The land acquisition for the project started in 2002 when notices were issued by the District Collector of Kalahandi to 12 villages in the Lanjigarh block to evacuate the land for the construction of the refinery (Das and Padel, 2020). Protests against the land acquisition were immediately initiated by local and professional activists and by the Kutia Kondhs (Borde and Bluemling, 2021). Only four out of 12 villages accepted the offer by the government, and the rest of the villages were left outside the boundary wall of the refinery (Kapoor, 2017; Das and Padel, 2020). The villages that rejected the offer formed the NSS in 2003 (Kapoor, 2017).

In 2003, Anil Agarwal rebranded Sterlite Industries as Vedanta Resources Ltd and listed the company on the London Stock Exchange (the first Indian to do so). Vedanta signed a new MoU with the government of Odisha that year to construct a one-million-tonnes-per-annum (MTPA) alumina refinery and a coal-based power plant in Lanjigarh. Vedanta also promised mining development at Lanjigarh in the Kalahandi district of Odisha (Temper and Martinez-Alier, 2013; Marshall and Balaton-Chrimes, 2016). Based on Vedanta's assurance of not diverting the forestland, the government of India's Ministry of Environment and Forests (MoEF) gave environmental clearance to Vedanta's project proposal in September 2004. Vedanta's Proposed Mining Lease (PML) also included proposals to mine bauxite from the Niyamgiri Mountains in the Raygada and Kalahandi districts in Odisha, an area adjoining the alumina refinery in Lanjigarh, which has a deposit of 73 million tonnes of mineable iron ore (Temper and Martinez-Alier, 2013; Marshall and Balaton-Chrimes, 2016).

The proposed mining site in the Niyamgiri Mountains is the home of the people from Dongria Kondh and Kutia Kondh tribes who are identified as Particularly Vulnerable Tribe Groups (PVTG) by the Indian government (Ministry of Tribal Affairs, 2019). According to the census data of 2001, the total population of the Dongria Kondhs is 7,952 (Saxena et al, 2010). This proposed project would have a detrimental impact on the environment and on the life and livelihood of the people living in that area. After a decade of continuous struggle, the NSS in solidarity with national and state activists was successful in getting a favourable decision from the Supreme Court of India in 2013 (see *Orissa Mining Corporation v Ministry of Environment and Forest and Others* 6 S.C.R. 881, 2013). In a precedent-setting judgment, the Supreme Court of India provided Gram Sabhas (village councils) with the power and authority to decide on the future of mining in the Niyamgiri Mountains (Marshall and Balaton-Chrimes, 2016). In the Gram Sabha proceedings carried out from July to August 2013, 12 villages unanimously voted against the mining activities in the Niyamgiri, essentially blocking Vedanta's path to bauxite extraction (Marshall and Balaton-Chrimes, 2016). Yet the on-ground mobilization continues in the face of further pushes for mining and intimidation from the state, even after the Supreme Court's landmark judgement in 2013.

In the face of this ongoing conflict where the permissions from local councils for land acquisition for mining is regularly bypassed and local resistance is threatened and crushed, we explore the evolving nature of the movement, paying attention to how collective action frames react to shifting political-economic contexts and the changing needs of the movement. These collective action frames act as a social and ideological adhesive to the movement, building solidarity among diverse movement participants even as the needs of the movement evolve. In the following discussion, we outline the content that upholds these frames, as well as the challenges faced by movement activists in adapting frames to shifting political-economic circumstances.

Building solidarity: articulating alternatives to development in movement framing

Tatpati et al (2016) suggest that there are limited academic analyses of the Dongria Kondhs' impression of development and wellbeing. In this study, we have focused on the ideas of development and wellbeing as conceptualized by the people of the Dongria Kondh community, together with those of popular intellectuals. This has allowed a broader understanding of the role of prefigurative politics within the movement and the inextricable connection between the health of the land and the health of the people. Jeffrey and Dyson (2021, pp 644–645) define 'prefigurative politics' as 'self-conscious channelling of energy into modelling the forms of action that are sought to be generalised in the future in circumstances characterised by power, hierarchy, and conflict'. Commitment to action against unjust political structures is one of the key components of prefigurative politics, and this form of politics upholds individual equality and freedom of expression (Jeffrey and Dyson, 2021). In this work, we have explored if such prefigurative articulations are included in the collective action framing of the Niyamgiri Movement.

A Dongria Kondh leader Mukesh,[5] for example, articulated the demands of NSS and his personal opinion on the issues of development: "If the doctor is called, he can't come immediately, he will come after two or three days. There should be a doctor in every *panchayat* [village council] in Niyamgiri. Why would we have to call the doctors from Raygada, Muniguda, Kalyansinghpur and Kalahandi?" (interview with Mukesh).

Mukesh clearly articulated demands for a better healthcare infrastructure in Niyamgiri rather than the government priority for wider roads of up to 30–40 feet. He felt that with the construction of roads, the Indigenous culture of the Dongria Kondhs would be threatened by newcomers. The Dongria Kondhs were happy with a five-feet road width (interview with Mukesh).

The issue of representation plays an important role in linking Indigenous articulations of development (Blaser, 2004). Nilsen (2016), while describing his experiences from the Narmada Bacchao Andolan (Save Narmada Movement, hereinafter NBA) observes that various actors within the NBA articulated the idea of development differently (see also Baviskar, 2004; Nilsen, 2010). A similar trend clearly emerged from our interviews. Activist Sunil believes that developmental policies pursued by the government are basically *bikash name binash* (destruction in the name of development). Sunil's active involvement with a range of other social movements in India might have shaped his vision of alternatives to development. Dongria Kondh leader Mukesh clearly articulated the immediate demands of the Dongria Kondhs, while Sunil, after describing these immediate demands, also spoke about establishing an 'economic democracy' (Kothari, 2019, p 67).

Economic democracy is one of the five spheres where 'local communities and individuals have control over the means of production, distribution, exchange and markets, based on the principle of localization for basic needs and trade built on this; central to this would be the replacement of private property by the commons, and increasing focus on the economy of caring and sharing' (Vikalp Sangam, n.d.). Kothari (2019) explains the five overlapping spheres of alternatives identified by the India-based organization Vikalp Sangam (Alternatives Confluence). Vikalp Sangam tries to bring together those organizations and movements that are working on alternatives (see Vikalp Sangam, n.d.). Sunil's vision of an alternative to the state's unilinear developmental model foregrounds 'situatedness' (Bebbington, 2004). Socioecological and agro-ecological situatedness of rural social movements as identified by Bebbington (2004) are prefigurative alternatives to corporate-led conceptualizations of development. Sunil proposes a place-based alternative to development with the participation of the tribal communities.

All the popular intellectuals involved with the Niyamgiri Movement clearly and unanimously identify the immediate demands of the Dongria Kondh community as paramount, but their vision of alternatives to development varies. There are some prefigurative ideas about alternative models of development among the movement activists, as mentioned by Sunil during his interview, but those ideas are not reflected in the existing framing of the movement. The existing framing continues to revolve around the ecological importance of Niyamgiri which played an important role in achieving a favourable judgement from the Supreme Court of India in 2013. The popular intellectuals involved with the movement have not gained much success in incorporating these evolving issues in the framing of the movement, which might have helped to galvanize solidarity both at the national and the international level. If popular intellectuals have not been successful in adapting the movement's frames to evolving alternate, articulations of development, what else might bind the collective action frame?

Education in Niyamgiri

The issue of education is intrinsically linked with the issue of development in Niyamgiri. During the interview with Dongria Kondh leader Mukesh, this issue of alternative education was raised, and his facial reactions changed to register despair. He explained that his children were living in a city near Bhubaneswar where they are continuing with their schooling, but he was upset because he had not met with them for an entire year; he did not know when he would see them next. As we spoke with Mukesh and the other activists of the NSS, most of them argued for an accessible education system in Niyamgiri and demanded education in their local language *Kui*.

Popular intellectual Ajit[6] stated that the NSS had repeatedly raised demands for school, education and issues such as health and sanitation with the state government, but nothing concrete had been done by any level of government. While this expectation of state delivery of social development was central in the discussion with popular intellectuals, in the case of the NSS, no alternatives to this framing were forthcoming; the views of Dongria Kondh and popular intellectuals were aligned. The Niyamgiri Movement has not come up with an alternative model for education. In fact, due to a lack of schools in Niyamgiri, parents were sending their children to private schools in other cities in Odisha. In the year 2019–2020 alone, 475 primary schools and 13 upper primary schools were closed in Odisha State (Xaxa, 2021). Activist Swaraj highlighted the challenge of the lack of educational facilities in Niyamgiri and its dire consequences for the future.

The Kalinga Institute of Social Science (KISS) located in Bhubaneswar, Odisha, is the world's largest private boarding school which claims to host 27,000 students from kindergarten to postgraduate level; all of them belong to the tribal communities. Since all the schools in Niyamgiri are closed, the Dongria children have to go to private boarding schools or to residential schools or *ashram* schools run by the government and private entities (Gupta and Padel, 2020). These private boarding schools like KISS are often funded by mining conglomerates (Xaxa, 2021). One media report (see BusinessLine, 2018) describes how Vedanta is sponsoring the education of tribal children at KISS. Gupta and Padel (2018) argue that these boarding schools are similar to North American 'industrial schools', which provided education to Native American children. This kind of education system supported by the extractive industries has been termed 'extraction education' (Walker, 2018, p 78) and is also visible in British Columbia, where the Liquid Natural Gas (LNG) industry engages in similar practices of steering hearts and minds through school provision.

When a tribal child is uprooted from their culture, their identity is lost and their traditional customs and rituals swept away by 'corporate values', which emphasize 'money and financial power' (Padel and Das, 2010, p 336). Dyson (2019) notes that the cultural trends of the dominant classes get recognised and rewarded in schools, and students from marginalized backgrounds feel further ignored and excluded. Gupta and Padel (2020) demonstrate how Hindu cultural and social values are systematically imposed on tribal (Adivasi) children at KISS. They draw examples from North America and Australia, where children were removed from their communities and placed into residential and boarding schools as a process of 'de-tribalizing', which they identify as a form of 'cultural genocide' (Gupta and Padel, 2019, p 82). Due to lack of education opportunities, there is a similar situation in Odisha, albeit the historical and sociopolitical context in India is different from that of Australia and North America. The tribal students in Odisha are being forced to go to residential private schools where they are systematically dispossessed from their land and livelihoods (Xaxa, 2021), which play a prevalent role in shaping their worldview.

However, elsewhere in India and beyond, there are many examples of social transformation through grassroots education. Social movements like the Zapatistas in Mexico and Narmada Bachao Andolan (NBA) in India developed alternative schooling systems to nourish alternative visions of development (see Nilsen, 2010; Routledge, 2017). NBA was successful in setting up an alternative schooling system known as Jeevanshalas ('The Life Schools') in two villages on the banks of the River Narmada (NBA Badwani, 2019a). These schools started in 1991–1992 and introduced an alternative curriculum emphasizing, for example, the skills of forest protection, soil conservation, watershed management and learning to use alternative sources of energy (Nilsen, 2010; NBA Badwani, 2019a). The Jeevanshalas remain operational today (seven Jeevanshalas in Maharashtra and two Jevansahalas in Madhya Pradesh) and most are remotely located where basic necessities of life are not easily accessible (NBA Badwani, 2019b). The Jevansahalas school curriculum tries to impart lessons on peaceful coexistence with nature, focusing on self-reliance and cooperation (NBA Badwani 2019b). The Zapatistas in Chiapas, Mexico, also reject all facilities provided by the state and run their own educational institutions, hospitals and healthcare system (Walsh, 2015; Oikonomakis, 2018). The community is self-sufficient in food production and the students in the Zapatista community get practical lessons outside the classroom to implement sustainable farming techniques (Walsh, 2015; Gahman, 2016). Despite education being a key concern among some popular intellectuals and the Dongria Kondh, this issue has not been incorporated into the collective action frame of the movement to mobilize an alternative form of development.

Suppressing solidarity: state oppression and constraints to collective actions frames

The previous two sections revealed how popular intellectuals have struggled to adapt the movement's collective action frames in terms of articulating alternatives to development and alternative education systems. The popular intellectuals involved have therefore not been completely successful at incorporating the pertinent issues of the Dongria Kondh into the existing collective action framing of the movement. For many of the popular intellectuals, this limitation lies not in a failed movement ideology or strategy, but in state-based oppression. Intimidation by the state continued to increase in Niyamgiri following the 2013 Supreme Court verdict. The central government of India continues its efforts to crush the Niyamgiri Movement by intentionally misidentifying the activists of NSS as Maoists.

For example, the central government of India tagged the NSS as a Maoist outfit in the 2017 annual report of the Ministry of Home Affairs (Ministry of Home Affairs, n.d., p 6). This identification by the government has been firmly refuted by the activists of NSS (Chakraborty, 2022).

From 1967 onwards, the Maoist movement in India has grown, spreading across ten states, including Odisha (Behera, 2020). According to Behera (2020), the Maoist insurgency in Odisha has increased since 2004. The geographical location of Odisha also played a significant role in the increase of Maoist activities; Odisha shares borders with states like Andhra Pradesh, Chhattisgarh and Jharkhand, in which there is a significant Maoist presence (Behera, 2020). The Maoist insurgents in Odisha began supporting movements against land grabbing and dispossession for major development projects which also included the people's resistance in Niyamgiri (Sahoo, 2019). They were successful in garnering support from the common people because they promised to address the pertinent issues of displacement, exploitation and deprivation (Behera, 2020). The Maoists have reached some remote pockets in India where state institutions have failed to hear the grievances of the Adivasi people, enabling Maoists to gain ground and support (Guha, 2007).

The perspective of the state on the issue of the Maoists largely differs from the view of the popular intellectuals involved with the Niyamgiri Movement. In 2009, the then Prime Minister of India, Dr Manmohan Singh, identified the Maoists as the 'gravest internal threat' (see *New Indian Express, 2009*) and the state continues to disavow them as a political force. On the other hand, most of the popular intellectuals identify the Maoists as a political force. Until and unless the state recognizes the Maoists and initiates a dialogue with them, it will be difficult for the movement activists to implement their alternative visions of development. During Chakraborty's informal conversations, several popular intellectuals involved with the Niyamgiri Movement mentioned finding it difficult to initiate a conversation with the Maoists because the Maoists claim that people are already in their control and fail to acknowledge other political forces. Further, even when the Maoists are invited for a dialogue by the popular intellectuals, they do not participate (Chakraborty, 2022).

The Central Reserve Police Force (CRPF) has occupied a panchayat (village council) office at the base of the Niyamgiri Mountains to conduct combing operations in Niyamgiri. Converting a local panchayat office into a CRPF camp has been condemned by the activists involved with the movement. Activist Ajit reaffirms the need to initiate a dialogue with the Maoists. Maoists have attacked schools, destroyed roads and other infrastructure projects in several parts of India where they are present (Behera, 2019; Shah, 2011). So, the state's initiation of an open dialogue with the Maoists and all the stakeholders will help the movement and the affected communities to articulate their vision of alternatives to development.

Conclusion

This chapter unravels the complexities of understanding place-based solidarity in framing the anti-mining movement in the Niyamgiri Mountains, Odisha, India. It has demonstrated these complexities by focusing on popular intellectuals who are innovators, movement intellectuals and movement allies. This chapter demonstrates areas where popular intellectuals have not been able to incorporate the key concerns of the Dongria Kondhs into the collective action framing of the movement; that is, the key issues of the Dongria Kondhs were not properly integrated. Since the collective action framing of the popular intellectuals is mostly centred on the sanctity of the Niyamgiri Mountains and its ecological significance, they were unsuccessful in incorporating issues to do with education development and healthcare infrastructure. Some external actors surely considered the 2013 Supreme Court verdict a success, but the Dongria Kondhs and the popular intellectuals know the real limitations of any success. Even though they were successful in stopping Vedanta from mining the mountains, many of the ongoing concerns of the Dongria Kondhs remain unheard and their connection to the Niyamgiri Mountains continue to be disavowed and undermined.

In addition, the intimidation by security forces has kept the movement activists and the popular intellectuals engaged in prolonged legal battles. This is also another reason why they have not been successful in incorporating their alternative visions of development into the collective action framing of the movement. As a result of this intimidation, the Niyamgiri Movement is in a state of self-defence where they are engaged in saving and protecting their key members from violence perpetuated by the state. The state has tactically kept the movement activists busy through its spectre of violence, which is preventing the formulation and incorporation of an alternative vision of development in the collective action framing of the Niyamgiri Movement. Irrespective of all the challenges encountered, the popular intellectuals involved in the Niyamgiri Movement have used their expertise and networks in building solidarity across scales, which needs to be acknowledged. The future success of the Niyamgiri Movement will depend on the incorporation of pertinent issues of development and the wellbeing of the Dongria Kondhs in the collective action framing of the movement. This will also help in galvanizing solidarity at the national and international levels.

Notes

1. Several authors have also used the spelling 'Dongaria Kondh' in their writings. In this chapter, we will be using the spelling 'Dongria Kondh'.
2. Like the Dongria Kondhs, Kutia Kondhs are a part of larger Kondh clan (Borde and Bluemling, 2021). Out of five sections of the Kondh tribe, the three main groups are Dongria Kondh, Kutia Kondh and Deshia Kondh. The Dongria Kondhs lives in the

mountains, the Kuita Kondhs lives in the foothills and the Deshia Kondhs are plainsmen (Jena et al, 2002, p 3).
3   Member of Lok Shakti Abhiyan and involved with National Alliance of People's Movement (NAPM).
4   The interviews were conducted in Hindi, English and Odia. The Odia interviews were translated with the help of Poushali Ghatak and Rupradha Banerjee Mitra, who are fluent Odia speakers not associated with the Niyamgiri Movement.
5   Active member of NSS.
6   Member of Samajwadi Jan Parishad (Socialist People's Council) and NSS.

References

Baud, M. and Rutten, R. (2005) 'Introduction', in M. Baud and R. Rutten (eds) *Popular Intellectuals and Social Movements: Framing Protest in Asia, Africa, and Latin America*, Cambridge: Cambridge University Press, pp 1–18.

Baviskar, A. (2004) *In the Belly of the River: Tribal Conflicts over Development in the Narmada Valley*, New Delhi: Oxford University Press.

Bebbington, A. (2004) 'Movements and modernizations, markets and municipalities', in R. Peet and M. Watts (eds) *Liberation Ecologies: Environment, Development and Social Movements*, London: Routledge, pp 358–382.

Behera, A. (2019) 'Politics of good governance and development in Maoist affected scheduled areas in India: a critical engagement', *Studies in Indian Politics*, 7(1): 44–55.

Behera, A. (2020) 'People's movement under a revolutionary brand: understanding the Maoist movement in Odisha', *Millennial Asia*, 11(2): 211–225.

Benford, R.D. and Snow, D. (2000) 'Framing processes and social movements: an overview and assessment', *Annual Review of Sociology*, 611–639, DOI: 10.2307/223459.

Blaser, M. (2004) '"Way of life" or "who decides": development Paraguayan Indigenism and the Yshiro people's life projects', in M. Blaser, H.A. Feit and G. McRae (eds) *In the Way of Development: Indigenous Peoples, Life Projects and Globalization*, London: Zed Books, pp 52–71.

Borde, R. and Bluemling, B. (2021) 'Representing Indigenous sacred land: the case of the Niyamgiri movement in India', *Capitalism Nature Socialism*, 32(1): 68–87.

Borde, R. and Rasch, E.D. (2018) 'Internationalized framing in social movements against mining in India and the Philippines', *Journal of Developing Societies*, 34(2): 195–218.

BusinessLine (2018) 'Vedanta funds education of 100 underprivileged tribal children', *The Hindu*, 12 March, Available from: https://www.thehindubusinessline.com/news/national/edanta-funds-education-of-100-underprivileged-tribal-children/article23080857.ece [Accessed 27 July 2021].

Chakraborty, S. L. (2022) 'Framing the pluriverse: understanding the role of popular intellectuals in the Niyamgiri Movement in India', Doctoral Thesis, Monash University Australia, Available from: https://bridges.monash.edu/articles/thesis/Framing_the_Pluriverse_Understanding_the_Role_of_Popular_Intellectuals_in_the_Niyamgiri_Movement_in_India/19769143 [Accessed 29 March 2024].

Das, S. and Padel, F. (2013) 'Battles over bauxite in East India: the Khondalite Mountains of Khondistan', in R.S. Gendron, M. Ingulstad and E. Storli (eds) *Aluminum Ore: The Political Economy of the Global Bauxite Industry*, British Columbia: UBC Press, pp 328–352.

Das, S. and Padel, F. (2020) *Out of This Earth: East India Adivasis and the Aluminium Cartel*, Telengana, India: Orient BlackSwan.

DeWalt, K.M. and DeWalt, B.R. (2011) *Participant Observation: A Guide for Fieldworkers*, Plymouth: AltaMira Press.

Dyson, J. (2019) 'Rethinking education as a contradictory resource: girls' education in the Indian Himalayas', *Geoforum*, 103: 66–74.

Eyerman, R. (1994) *Between Culture and Politics: Intellectuals in Modern Society*, Cambridge: Polity Press.

Gahman, L. (2016) 'Food sovereignty in rebellion: decolonization, autonomy, gender equity and the Zapatista solution', *Solutions*, 7(4): 67–83.

Government of Odisha (2021) 'Odisha Economic Survey 2020–21. Bhubaneswar', Available from: https://finance.odisha.gov.in/sites/default/files/2021-02/Economic_Survey.pdf [Accessed 7 March 2024].

Guha, R. (2007) 'Adivasis, Naxalites and Indian democracy', *Economic and Political Weekly*, 42(32): 3305–3312.

Gupta, M. and Padel, F. (2018) 'Confronting a pedagogy of assimilation: yhe evolution of large-scale schools for tribal children in India', *JASO-online*, X(1), Available from: https://www.anthro.ox.ac.uk/files/jaso1012018pdf [Accessed 27 July 2021].

Gupta, M. and Padel, F. (2019) 'Indigenous knowledge and value systems in India: holistic analysis of tribal education and the challenge of decentralising control', in M.C. Behera (ed) *Shifting Perspectives in Tribal Studies*, Itanagar: Springer Singapore, pp 67–86.

Gupta, M. and Padel, F. (2020) 'The travesties of India's tribal boarding schools', *Alice News*, Available from: https://alicenews.ces.uc.pt/?lang=1&id=31800 [Accessed 27 July 2021].

Jeffrey, C. and Dyson, J. (2021) 'Geographies of the future: prefigurative politics', *Progress in Human Geography*, 45(4): 641–658.

Jena, M.K. et al (2002) *Forest Tribes of Orissa: The Dongaria Kondh*, New Delhi: D.K. Prtintworld (P) Ltd.

Kale, S.S. (2013) 'Democracy and the state in globalizing India: a case study of Odisha', *India Review*, 12(4): 245–259.

Kapoor, D. (2017) 'Adivasi, Dalit, and Non Tribal Forest Dweller (ADNTFD) resistance to bauxite mining in Niyamgiri: displacing capital and state corporate mining activism in India', in D. Kapoor (ed) *Against Colonization and Rural Dispossession: Local Resistance in South and East Asia, the Pacific and Africa*, London: Zed Books, pp 67–97.

Kothari, A. (2019) 'Radical well-being alternatives to development', in P. Cullet and S. Koonan (eds) *Research Handbook on Law, Environment and the Global South*, Chelthenham: Edward Elgar, pp 64–84.

Krishnan, R. and Naga, R. (2017) '"Ecological warriors" versus "indigenous performers": understanding state responses to resistance movements in Jagatsinghpur and Niyamgiri in Odisha', *South Asia: Journal of South Asia Studies*, 40(4): 878–894.

Kumar, K. (2014) 'The sacred mountain: confronting global capital at Niyamgiri', *Geoforum*, 54: 196–206.

Lahiri-Dutt, K. (2014) 'Introduction to coal in India: energising the nation', in K. Lahiri-Dutt (ed) *The Coal Nation Histories, Ecologies and Politics of Coal in India*, Farnham: Ashgate, pp 1–35.

Macdonald, K., Marshall, S. and Balaton-Chrimes, S. (2017) 'Demanding rights in company-community resource extraction conflicts: examining the cases of Vedanta and POSCO in Odisha, India', in J. Grugel et al (eds) *Demanding Justice in The Global South: Claiming Rights*, Cham: Springer International Publishing, pp 43–67.

Marshall, S.D. and Balaton-Chrimes, S. (2016) 'Tribal claims against the Vedanta bauxite mine in Niyamgiri, India: what role did the UK OECD national contact point play in instigating free, prior and informed consent?', Non-judicial Redress Mechanisms Report Series 9, Available from: https://papers.ssrn.com/sol3/papers.cfm?abstract_id=2878211 [Accessed 12 March 2024].

Ministry of Home Affairs (n.d.). *Annual Report 2016–17, Government of India*, Available from: https://www.mha.gov.in/sites/default/files/AnnualReport_16_17.pdf [Accessed 12 March 2024].

Ministry of Tribal Affairs (2019) 'Welfare of particularly vulnerable tribal groups', Press Information Bureau, Government of India, Available from: https://pib.gov.in/Pressreleaseshare.aspx?PRID=1577166 [Accessed 13 January 2022].

NBA Badwani (2019a) 'Jeevanshala – the "life schools"', Available from: https://narmadaandolan.org/jeevanshala-the-life-schools/ [Accessed 30 September 2021].

NBA Badwani (2019b) 'Jeevanshalas today', Available from: https://narmadaandolan.org/jeevanshalas-today/ [Accessed 30 September 2021].

New Indian Express (2009) 'Maoists gravest threat to security: PM', Available from: https://www.newindianexpress.com/nation/2009/sep/15/maoists-gravest-threat-to-security-pm-86582.html [Accessed 13 August 2021].

Nilsen, A.G. (2010) *Dispossession and Resistance in India: The River and the Rage,* New York: Routledge.

Nilsen, A.G. (2016) 'Power, resistance and development in the Global South: notes towards a critical research agenda', *International Journal of Politics, Culture and Society,* 29(3): 269–287.

Oikonomakis, L. (2018) *Political Strategies and Social Movements in Latin America: The Zapatistas and Bolivian Cocaleros,* London: Palgrave Macmillan.

Orissa Mining Corporation v Ministry of Environment and Forest and Others 6 S.C.R. 881 (2013) Available from: https://indiankanoon.org/doc/109648742/ [Accessed 29 March 2024].

Oskarsson, P. (2013) 'Dispossession by confusion from mineral-rich lands in Central India', *South Asia: Journal of South Asian Studies,* 36(2): 199–212.

Padel, F. and Das, S. (2007) '"Agya, what do you mean by development?"', in R. Kalshian (ed) *Caterpillar and the Mahua Flower: Tremors in India's Mining Fields,* New Delhi: PANOS South Asia, pp 24–46.

Padel, F. and Das, S. (2010) 'Cultural genocide and the rhetoric of sustainable mining in East India', *Contemporary South Asia,* 18(3): 333–341.

Padhi, R. and Sadangi, N. (2020) *Resisting Dispossession: The Odisha Story,* London: Palgrave Macmillan.

Pattnaik, B.K. (2013) 'Tribal resistance movements and the politics of development-induced displacement in contemporary Orissa', *Social Change,* 43(1): 53–78.

Perreault, T. (2013) 'Dispossession by accumulation? Mining, water and the nature of enclosure on the Bolivian Altiplano', *Antipode,* 45(5): 1050–1069.

Routledge, P. (2017) *Space Invaders: Radical Geographies of Protest,* London: Pluto Press.

Rutten, R. and Baud, M. (2005) 'Concluding remarks: framing protest in Asia, Africa, and Latin America', in M. Baud and R. Rutten (eds) *International Review of Social History,* Cambridge: Cambridge University Press, pp 197–217.

Sahoo, N. (2019) 'Half a century of India's Maoist insurgency: an appraisal of state response', Observer Research Foundation, Occasional Paper, Available from: https://www.orfonline.org/research/half-a-century-of-indias-maoist-insurgency-an-appraisal-of-state-response-51933/ [Accessed 28 September 2021].

Sahu, G. (2019) 'Forest rights and tribals in mineral rich areas of India: the Vedanta case and beyond', in P. Cullet and S. Koonan (eds) *Research Handbook on Law, Environment and the Global South,* Cheltenham: Edward Elgar, pp 272–285.

Saxena, N.C. et al (2010) *Report of the Four Member Committee for Investigation in the Proposal Submitted by the Orissa Mining Company for Bauxite Mining in Niyamgiri,* Ministry of the Environment and Forests, Government of India.

Shah, A. (2011) 'India burning: the Maoist revolution', in I. Clark-Decès (ed) *A Companion to the Anthropology of India*, Chichester: Wiley-Blackwell, pp 332–351.

Snow, D. (2004) 'Framing processes, ideology, and discursive fields', in D. Snow, S.A. Soule and H. Kreisi (eds) *The Blackwell Companion to Social Movements*, Oxford: Blackwell, pp 380–412.

Snow, D. and Benford, R.D. (1988) 'Ideology, frame resonance and participant mobilization', *International Social Movement Research*, 1: 197–217.

Snow, D., Vliegenthart, R. and Ketelaars, P. (2019) 'The framing perspective on social movements', in D.A. Snow, S.A. Soule, H. Kreisi, & H.J. McCammon (eds), *The Wiley Blackwell Companion to Social Movements* (2nd edn), Chichester: John Wiley & Sons, pp 392–410.

Snow, D. et al (1986) 'Frame alignment processes, micromobilization, and movement participation', *American Sociological Review*, 51(4): 464–481.

Tatpati, M., Kothari, A. and Mishra, R. (2016) *The Niyamgiri Story: Challenging the Idea of Growth without Limits?* Pune, Maharashtra: Kalpavriksh, Available from: https://kalpavriksh.org/wp-content/uploads/2018/06/NiyamgiricasestudyJuly2016.pdf [Accessed 12 March 2024].

Temper, L. and Martinez-Alier, J. (2013) 'The god of the mountain and Godavarman: Net Present Value, indigenous territorial rights and sacredness in a bauxite mining conflict in India', *Ecological Economics*, 96: 79–87.

Vikalp Sangam (n.d.) 'About Vikalp Sangam', Available from: https://vikalpsangam.org/about/ [Accessed: 28 September 2021].

Walker, J. (2018) 'Creating an LNG ready worker: British Columbia's blueprint for extraction education', *Globalisation, Societies and Education*, 16(1): 78–92.

Walsh, C.E. (2015) 'Decolonial pedagogies walking and asking. Notes to Paulo Freire from AbyaYala', *International Journal of Lifelong Education*, 34(1): 9–21.

Worsdell, T. and Sambhav, K. (2020) *Locating the Breach: Mapping the Nature of Land Conflicts in India*, New Delhi: Land Conflict Watch, Available from: https://d1ns4ht6ytuzzo.cloudfront.net/oxfamdata/oxfamdatapublic/2020-02/Locating the Breach Report-Final.pdf [Accessed 31 December 2022].

Xaxa, V. (2021) 'How KISS and Indian anthropology degrade tribal people', *The Caravan*, Available from: https://caravanmagazine.in/education/kiss-kalinga-tribes-adivasi-anthropology-world-congress [Accessed 3 August 2021].

8

Solidarity as Praxis in Class Struggle

Laura Bedford

Without a fight, there's no environmental justice, there's no victory. We will continue fighting and we will continue defending our territory and our lives and the lives of our children and the lives of future generations. We are fighting not only for our community, but for yours too – for food sovereignty, for access to water free from contamination and for a healthy environment. We need you to join us in this fight.
Eliete Paraguassu, shellfish gatherer and defender of her community against the occupation by petroleum companies, Ilha de Maré, Brazil in Global Witness, 2022, p 30

In light of slow violence (Nixon, 2011) being faced by land and environmental defenders around the planet – including Eliete Paraguassu (quoted above), a Quilombola fisher whose livelihood is threatened by oil extractivism in the previously pristine island bay of Maré Island she calls home – how might people in positions of relative privilege, but who recognize the systemic violence facing people and the planet, act in solidarity to engage with this violence and its impunity? How might we 'join the fight'? While mainstream climate activists in the global economic core argue in vague terms that the 'economy' is the problem, criticize 'growth' and claim that 'we need to keep the carbon in the ground', any articulation of class struggle or a decolonial future which envisages a redistribution of economic and political power is largely absent from their account. In this chapter, I argue that without reference to class solidarity in the fight against capitalism and colonialism, discussions of planetary justice are moot and leave us with an impuissant and vague environmentalism which is no longer fit for purpose.

Like all systems of governance under capitalism going back to the sixteenth century, the rise of mass industry and chattel slavery, current institutions and instruments established to chart a way out of fossil-fuel dependence – including those under the auspices of the United Nations (UN) – continue their original intent to support capitalist growth and profit. International and national policies get designed and redesigned to strengthen and protect capitalist hegemony and the ruling elite. Under the auspices of the post-Second World War Bretton Woods institutions and the UN, with its smattering of representation of the world's people in the form of compradors and plutocrats, we are as a planet where we are now. Under the Washington Consensus, the capitalist 'development' paradigm adopted by the UN has peddled the myth that global inequality stems from inclusion and exclusion from capitalist markets when, instead, poverty is globally sustained by class relations shaped by states through policies that benefit capital, especially finance capital (Selwyn, 2014). Today the richest *eight* people control as much wealth as the poorest half of the planet's population combined, and of the ten wealthiest individuals in the world, all white men, only one lives outside the US (Moskowitz, 2023). How might we begin to work towards planetary justice given this extreme inequality, and what is the meaning of solidarity?

Capitalism requires endless economic growth, and this leaves the 'green capitalist' climate movement in the economic core with a choice to make: either it 'ignores the ecological devastation wreaked by corporations, or it commits itself to tackling the root cause of that devastation. If it is truly internationalist and truly planetary, it must recognise that climate struggle is class struggle. If it values all bodies equally, it must denounce capitalism itself' (Dhillon, 2019, para 16). There is a creeping tendency in much environmental discourse towards neo-Malthusianism and eugenics dressed up as ecological justice and climate action, which need to be reckoned with. This includes efforts to privatize the remaining commons and 'natural capital', to safeguard it from the poor (see, for example, the arguments of influential Cambridge economics Professor Partha Dasgupta [2021]). As Nnimmo Bassey has stressed, so-called 'green capitalism' or 'market environmentalism' advances the view that the:

> basis for nature can only be preserved when it is assigned monetary value. This position is sold as green economy and fits well into neoliberal constructs – presenting speculators with opportunities to reap profits from ecological destruction originating from, but not limited to, extractivism, land grabs, genetic manipulations and a number of techno-fixes that ensure the reign of monopolies. (Bassey, 2013, para 4)

If we are sufficiently reflexive, we will realize, as criminologist Biko Agozino argues, that '[d]ecolonization requires alliances and coalitions in articulation.

There are definitely comparable facts about the colonizer and the colonized society when it comes to the police, courts, and prisons, war, capitalism, racism, sexism, and struggles against the ills of society that make up the culture of resistance' (2023, p 439). Indeed, 'decolonization is not a struggle left only to the colonized, it is a struggle open to contributions from all angles' (Agozino, 2023, p 444). To overthrow capitalism and colonialism, we need to actively seek to understand, even as privileged outsiders, the struggles and triumphs of oppressed and exploited people, and work to erode the capitalist realism (Fisher, 2009), which is so glibly peddled by our national and global institutions of governance, and that make this oppression and class inequality seem natural and inevitable. This is the meaning of solidarity. So too is support for decolonization and reparations as part of the reckoning with the ongoing damage of colonialism and slavery.

However, at a global level, there is now a 'widespread sense that not only is capitalism the only viable political and economic system, but also that it is now impossible even to imagine a coherent alternative to it' (Fisher, 2009, p 2). For this reason, as Foster et al (2020, p 2) argue, '[r]eturning to Marx as a *starting point* is crucial in order to develop a materialist critique of capitalism and colonialism'. This stands in sharp contrast to postmodern leftish sentimentality and its rejection of truth or objective reality – and the ultimate contradiction that this leads to an embrace of the metanarratives of capitalist realism. Rather than offering a reductive, globalizing, deterministic, Promethean vision of a postcapitalist alternative, Foster et al argue that in Marxism:

> there is no such thing in historical materialism as a fixed orthodoxy. Rather, Marxism from the beginning has been shaped by vernacular revolutionary traditions. As a philosophy of praxis geared not simply to understanding the world but also changing it, historical materialism can least of all afford to be supra-historical or to neglect the lessons of national and popular struggles. (Foster et al, 2020, p 2)

Further, as global systems theorist and Marxist scholar Samir Amin points out, 'Marxism does not reduce social reality to economic determinism. In fact, the economic reductionism of contemporary thought is expressed in the everyday (naïve) language of governments, every time that they formulate economic constraints as a "laws of nature"!' (Amin, 1992, p 527). Foster et al argue that orthodox Marxism refers exclusively to method and it 'is thus the materialist, historical, and dialectical method of classical Marxism that constitutes the necessary point of departure with which to engage in the critique of colonialism, including settler colonialism, today' (Foster et al, 2020, p 2). And, of course, a Marxist analysis comes back to understanding the role of class solidarity in confronting capitalism (Marx and Engels, 2018 [1848]).

So for those of us in privileged positions in the global economic core who are wondering where to take aim, the least we can do is to work to understand – to know, learn, unlearn and seek the truth – and act in solidarity with those who are today and have been for hundreds of years resisting capitalism, colonialism, enclosure, primitive accumulation, extraction, racism, ecocide and genocide, and who stand as a last bulwark between capitalists and planetary catastrophe. It is not about tokenism, and the optics of the 'diversification' of a white politics of protest, as those in the mainstream climate movement have scrambled to 'correct' (see, for example, Extinction Rebellion, 2023). Neither is it a retreat into tactics of 'deference politics' that 'insulate us from criticism [but] also insulate us from connection and transformation. They prevent us from engaging empathetically and authentically with the struggles of other people – prerequisites of coalitional politics' (Táíwò, 2020, p 66). Instead, it is 'crucial that the work of land and environmental defenders is continued and amplified' (Global Witness, 2022, p 29). As Táíwò argues, a constructive approach to 'standpoint epistemologies' would 'calibrate itself directly to the task of redistributing social resources and power rather than to intermediary goals cashed out in terms of pedestals or symbolism' and it would seek to counter the erosion of the 'practical and material bases for popular power over knowledge production and distribution, particularly that which could aid effective political action and constrain or eliminate predation by elites' (Táíwò, 2020, p 67).

Given the ongoing primitive accumulation in the periphery as we move further into the 21st century – and the attendant ecocide and genocide – the reality of the harms of the international capitalist system is best understood by those on the frontline of the ecological war, and those enmeshed within social and ecological damage of state-corporate 'regimes of permission' (Whyte, 2014, p 237). Those already aware of their oppression as a class are also best placed to lead the struggle because the growth of 'awareness of the objectives of the [class] struggle and the struggle itself, are not different things separated chronologically and mechanically. They are only different aspects of the same struggle' (Luxemburg, 1961 [1904], p 88). Indeed, 'praxis is the process of the formation of class consciousness, and it is precisely the exploited classes that will be at the forefront, unveiling the façade that capitalism is the natural and inevitable state of social relations of production and reproduction' (Bedford, 2022, p 216).

Vandana Shiva (*Global Witness*, 2022, p 6) reminds us that 'around the world, three people are killed every week while trying to protect their land, their environment, from extractive forces'. And 'corporate power, supported by government policies, is a significant underlying force that has not only driven the climate and biodiversity crisis to the brink, but which has continued to perpetuate the killing of defenders' (*Global Witness*, 2022, p 26). In 2021 alone, 200 defenders were murdered. Of course, to act in solidarity with land and environmental defenders in the periphery and semi-periphery of the economic system means that all of us who are able to do so should continue to 'shine a light on the stories of resistance and repression, not just to remember those who have been killed in the line of resistance but to continue their urgent work by telling the world exactly why they are dead' (Shiva, in *Global Witness*, 2022, p 6) . We must continue to 'honour the dead with [our] attention. To get angry on their behalf, and then to act' (Shiva, in *Global Witness*, 2022, p 6). In the decade since *Global Witness* started reporting on the killing of environmental and land defenders, 39 per cent of the 1,733 killed were Indigenous people, which is likely a very serious undercount, given the strict requirements for verification that *Global Witness* uses (*Global Witness*, 2022, p 16). Whereas 'both states and corporations have directly instrumentalized their own security organizations to exert deadly repression, notably in the context of public protests, more insidious forms of repression – including targeted killing – are generally conducted through intermediaries including former military officers, private security contractors, and criminal entrepreneurs (Middeldorp and Le Billon, 2019, pp 5–6). These deaths are seldom prosecuted and constitute the acute face of the slow and systemic violence individual defenders and communities face all around the world on a daily basis.

In settler colonies such as Australia, solidarity in the fight against the causes of climate change requires acknowledgement that:

> racialized and class-based functions of carceral power today ... [and] [e]ven where racism is not the guiding force for sustaining class divisions through criminalization, class can be significant ... this does not preclude analyses that seek to show the connections between materialist interests and domination. The production of racialized deprecating tropes ... is frequently central to reproducing systems of labour exploitation. (Ciocchini and Greener, 2021, p 1619)

With reference to the slow violence of settler colonialism specifically, Coulthard is clear that 'settler-colonialism should not be seen as deriving its reproductive force solely from its strictly repressive or violent features, but rather from its ability to produce forms of life that make settler-colonialism's constitutive hierarchies seem natural' (2014, p 152).

In light of this, we also need to unpack what we understand by violence and why we, in the economic core, feel we should be protected from it. The violence of capitalism and colonialism towards people and planet is not in question and has been well documented, even back to Marx, who highlighted how the 'profound hypocrisy and inherent barbarism of bourgeois civilization lies unveiled before our eyes, turning from its home, where it assumes respectable forms, to the colonies, where it goes naked' (Marx [1853], in Simpson, 1973, p 423).

I argue that given the unprecedented levels of systemic and structural violence against people and the planet that have come to be normalized in our social relations of production in the world and their entrenching into climate policy, a radical systems change is needed, not political promises and economic reforms from racist, capitalist state managers beholden to finance capital. Indeed, for 'Indigenous nations to live, capitalism must die. And for capitalism to die, we must actively participate in the construction of Indigenous alternatives to it' (Coulthard, 2014, p 173). A climate movement that expects anything more than the heretofore litany of misfeasance and malfeasance from the global so-called leadership and their instruments – a corrupt international plutarchy that has achieved *absolutely no progress* in mitigation of species extinction, genocide, ecocide or greenhouse gas emissions in the past 30 years – cannot possibly be expected to advance a revolutionary overthrow of our current economic system in the face of climate change. Indeed, 'extreme inequalities, environmental destruction and vulnerability to crisis are not a flaw in the system, but a feature of it. Only large-scale systemic change can resolve this dire situation' (UNRISD, 2022, p 47).

The climate movement in the economic core continues to reinforce the messaging through curated peaceful protest *against climate change*: that the compradors of global international capital will 'fix' the problem *if only* they would 'listen to the science', care for *their* children and somehow develop the political will to prioritize people and the planet over money and cut CO_2 pollution. Solidarity is, however, recognizing that capitalism 'is not a spontaneous order ... This creation of capitalism is the violence "securing insecurity"; it is the process of pacification ... [and] [i]t is pacification as a political technology for organizing everyday life through the production and re-organization of the ideal citizen-subjects of capitalism' (Neocleous, 2011, p 192). While Greta Thunberg has bravely argued that the only thing that creates hope is action (Thunberg, 2019), my question for her and the movement she has inspired in the economic core is: 'what is this action for or against?' And what is meant by action? By focusing energies on the international plutarchy and expecting them to *act* to reduce greenhouse emissions, is it possible that the 'climate movement' is steering the struggle off-course?

Anticipating that the capitalist system will prevent the worst excesses of global heating is risky. Capitalists will not change course unless they can do so in ways that allow them to 'externalize' the impacts, 'pivoting' their plunder to protect their profits. Neither will they accept policy imperatives that compel them to do so. International and state policy apparatus are not up to the task as they currently stand. As Foster argues, over the past half-century, under neoliberalism, 'the state is selectively withered away in its relation to capital, confined by its own self-imposed rational-legal character that must conform to the formal economic laws of the capitalist system, of which it is, paradoxically, the main legitimating force and official guarantor' (2019, p 9). Further, he notes that these limitations become starkly apparent 'whenever a social democratic government is brought to power, thinking it can institute reforms, only to discover that it is compelled to enforce neoliberal policies' (2019, p 9). Yet the neoliberal capture of social domains previously outside of the market has expanded the terrain of struggle, 'creating the wider terrains of class, race, social-reproductive, and environmental struggle, which today are merging to a remarkable degree in response to neoliberal absolute capitalism' (Foster, 2019, p 7).

The working class remains the 'biggest threat to capital today' and this 'is true both in the advanced capitalist countries themselves and even more so in the periphery, where the working class overlaps with the dispossessed peasantry. The working class is most powerful when able to combine with other subaltern classes as part of a hegemonic bloc led by workers' (Foster, 2019, p 6). The challenge is to overcome the mediated, constructed and entrenched divisions in relation to gender, race, class and religion; to see class as central. Certainly, as Canadian sociologist and unionist Kevin MacKay argues:

> Radically transforming industrial, capitalist civilization won't be easy. It will require movements for environmental sustainability, social justice, and economic fairness to come together, and to realize their common interest in dismantling the system of oligarchy and building a democratic, eco-socialist society. This 'movement of movements' must put aside sectarian squabbles, and finally realize that the goals of economic justice, human rights, and ecological sustainability are all intrinsically linked. (MacKay, 2018, para 18)

Young people in each generation have always led the way when it comes to delimiting the boundaries of political activism and resistance tactics. In a break with this historical role of young people, I argue, highly curated climate protests in the economic core may serve to relieve the worst symptoms of the cognitive dissonance the protesters feel, but do they constitute action for change? When young people direct their anger to the spectre of climate change and so-called 'corporate greed' – corporations are fine, just not *greedy* ones – the meaning of activism and resistance as revolutionary praxis becomes misdirected. This is because rather than seeing themselves as resisting oppression or acting in solidarity with oppressed groups, they only imagine themselves as needing to be heard. As Paulo Freire reminds us, the oppressed 'whose task it is to struggle for their liberation together with those who show true solidarity, must acquire a critical awareness of oppression through the praxis of this struggle' (2005 [1970], p 51). To overcome oppression and 'turn upon it ... can be done only by means of the praxis: reflection and action upon the world in order to transform it' (Freire, 2005, p 51). Acting on climate change is like shadow boxing. Climate change is a symptom of a sick system; it cannot be fought in any way other than directly challenging the system that gives rise to it. Even though reducing atmospheric carbon pollution at the speed required to prevent planetary catastrophe is increasingly unlikely, planetary justice is impossible without understanding the nature of the fight.

The nature of the fight will only be realized through global solidarity and the praxis of radical change making. Given the doublespeak and hypocrisy of our current so-called leadership, at the very least, the citizens – and certainly activists – in counties in the economic core that benefit from past and present genocide, ecocide and theft on an unimaginable scale around the world can do is acknowledge this past *and its present* manifestations. Beyond this, if they wish to be part of the revolutionary change, they will need to act in the interests of the working class, landless peasants, subsistence farmers and Indigenous peoples – against capital. In so doing, they may reveal their own status as members of an oppressed class. Environmental and climate movements in the economic core will – unless they act in solidarity with class-based movements and those struggling on the ecological frontline against enclosure, environmental racism and repression – make themselves effectively irrelevant, even though their activism garners undue media attention at home due to their privileged position and white supremacy.

Ordinary people in the economic core and peripheries are increasingly aware of the cause of their exploitation and that of the planet. Wherever they are in the world, in villages in the Sundarbans, townships and informal settlements in South Africa or New York City, people, acting in solidarity, can – if they are acting collectively – challenge capital. Yet to do so, we must name and act upon the system of oppression clearly, not cravenly gesture towards climate change. Indeed, 'human agents can of course change their social worlds for the better, as long as they are actually acting on real underlying forces, processes and structures rather than simply gesturing toward them' (Winlow and Hall, 2016, p 88). The important thing then, about praxis is that it requires action directed at the real cause of alienation and exploitation, not *causes* that are socially constructed. However, it is only through the experiential process of revolutionary praxis that oppressed and exploited people can change *both* their conditions of exploitation and develop *class* consciousness – that is, realize that they are oppressed *as a class*. The one change does not precede the other (Bedford, 2022). At the same time, as Kovel argues:

> Eco-socialism is no more a purely economic matter than was socialism or communism in the eyes of Marx. It needs to be precisely the radical transformation of society – and human existence – that Marx envisioned as the next stage in human evolution. Indeed, it must be that if we are going to survive the ecological crisis. (2007, p 1)

The 'climate movement'-mediated protests that have been established to 'fight climate change' aim to raise awareness of impending catastrophe and demand that politicians respond to 'citizens assemblies' to use their influence and policy to make capitalists in the economic core more responsive to the climate crisis. Yet the movement is at an impasse unless it acts *in solidarity* with movements that are aimed at decolonization and dismantling state repression. A climate movement that does not seek to highlight the systemic and structural antecedents of the climate crisis – capitalism – is missing the point. Capitalism cannot be reformed or degrown through policy mechanisms; it can only shape shift. Our current political systems, both national and international, under capitalism are not democracies but plutarchies. Imperialism is integral to the international capitalist mode of production, and decolonization is therefore not possible under global capitalism. It is not up to any self-proclaimed 'climate movement' to define the boundary of ethical tactics, especially a movement that eschews political discussion and debate and holds nonviolence (of the privileged) sacred. Instead, people themselves will mobilize based on their real material circumstances in myriad ways that suit their material conditions.

While on the one hand in the economic core, there is limited solidarity with the international working class or peasantry and 'it is clear that the current struggles in the West are occurring without any interest in what is happening elsewhere in the world … This is not the least of the successes of the dictatorship of the oligarchies and the use they make of their media clergy' (Amin, 2016, p 8). As Gramscian scholar Peter Thomas notes, 'without the elaboration of its own philosophy, or concrete conception of the world elaborated in institutions adequate to the specificity of its own project, the movement of the oppressed and exploited classes of capitalist society will remain subaltern to the existing dominant conception of the world' (2009, p 452). On the other hand, the oppressed classes are already mobilizing on the frontlines of the *ecological war* around the world through movements that challenge racist oppression and state violence, such as millions marching under the banner Black Lives Matter, and through union mobilization such as the marches of millions taking place across the UK and Europe in 2023 protesting against the erosions of workers' rights and social benefits. These protests are much larger and have broader bases than those that have characterized the climate movement in the economic core, despite claims by the green capitalist environmental movement to mass mobilization. In the act of mobilizing to resist against the common enemy – capitalism – people will drive forward deep transformation, but it will not take the form of a 'climate movement'. A multiplicity of tactics should be welcome, but the specific tactics will be determined by the exploited classes themselves, the landed and landless peasants, Indigenous peoples, the wage slaves, the unemployed, underemployed and precariously employed working class.

So long as the market and capital dictate global policy and our planetary next steps, global solidarity and planetary justice requires fight; it requires struggle and this cannot be left up to those who are least complicit in global warming to fight alone. Instead, now is the time that those of us in privileged spaces in the economic core try much harder to find ways to join the fight to bring the planet back into care and reciprocity. It is after all our responsibility to do so. Those with resources might use the courts and the co-opted 'democratic' party system as a first step towards reform, and to wrest back the power of the union movement. But the risk here is system relegitimization (Bond, 2013, p 59). Some might decide to act with others to collectively withhold their labour from those who seek to exploit it in order to generate toxic growth; some might disrupt and blockade extractives activities; some might wrest control of the means of production on the land from the oil companies, banks and agrochemical corporations to protect and expand alternative economic and agricultural systems; some might focus on survivance and resurgence (Simpson, 2014). And yes, some might blow up pipelines (Malm, 2021).

References

Agozino, M. (2023) 'The decolonization paradigm in criminology', in C. Cunneen, A. Deckert, A. Porter, J. Tauri and R. Webb (eds) *The Routledge International Handbook on Decolonizing Justice*, Abingdon: Taylor & Francis, pp 437–447.

Amin, S. (1992) 'Can environmental problems be subject to economic calculations?', *World Development*, 20(4): 523–530.

Amin, S. (2016) 'Reading capital, reading historical capitalisms', *Monthly Review*, 68(3): 1–18.

Bassey, N. (2013) 'Between Eti Uwem and green capitalism (green democracy)', *Africavenir*, AfricAvenir International, February, Available from: https://www.africavenir.org/news-details/archive/2013/february/article/nnimmo-bassey-between-eti-uwem-and-green-capitalism-green-democracy.html [Accessed 20 July 2023].

Bedford, L. (2022) 'In defense of class struggle', *Critical Criminology*, 30: 213–223, DOI: 10.1007/s10612-021-09567-z.

Bond, P. (2013) 'Climate crisis, carbon market failure, and market booster failure: a reply to Robin Hahnel's "Desperately Seeking Left Unity on International Climate Policy"', *Capitalism Nature Socialism*, 24(1): 54–61, DOI: 10.1080/10455752.2012.759364.

Ciocchini, P. and Greener, J. (2021) 'Mapping the pains of neo-colonialism: a critical elaboration of southern criminology', *British Journal of Criminology*, 61(6): 1612–1629.

Coulthard, G.S. (2014) *Red Skin, White Masks: Rejecting the Colonial Politics of Recognition*, Minneapolis: University of Minnesota Press.

Dasgupta, P. (2021) *Final Report – The Economics of Biodiversity: The Dasgupta Review*, HM Treasury, 2 February, Available from: https://www.gov.uk/government/publications/final-report-the-economics-of-biodiversity-the-dasgupta-review [Accessed 19 July 2023].

Dhillon, A. (2019) 'Extinction Rebellion must decide if it is anti-capitalist – and this greenwashing mining company shows us why', *The Independent*, 23 October, Available from: https://www.independent.co.uk/voices/extinction-rebellion-climate-crisis-bhp-mining-coal-colombia-a9167601.html [Accessed 19 July 2023].

Extinction Rebellion (2023) 'About us', Available from: https://rebellion.global/about-us/ [Accessed 19 July 2023].

Fisher, M. (2009) *Capitalist Realism: Is There No Alternative?*, Winchester: John Hunt Publishing.

Foster, J.B. (2019) 'Absolute capitalism', *Monthly Review*, 71(1): 1–9.

Foster, J.B., Clark, B. and Holleman, H. (2020) 'Marx and the Indigenous', *Monthly Review*, 71(9): 1–19.

Freire, P. ([1970] 2005) *Pedagogy of the Oppressed* (30th anniversary edition), New York: Continuum.

Global Witness (2022) 'Decade of defiance: ten years of reporting land and environmental activism worldwide', *Global Witness*, 29 September, Available from: https://www.globalwitness.org/en/campaigns/environmental-activists/decade-defiance/ [Accessed 19 July 2023].

Kovel, J. (2007) 'Why ecosocialism today?', *New Socialist*, 61(10), Available from: https://climateandcapitalism.com/2007/07/24/joel-kovel-why-ecosocialism-today/ [Accessed 19 July 2023].

Luxemburg, R. ([1904] 1961) *The Russian Revolution, and Leninism or Marxism?*, B. Wolfe (trans) Ann Arbor: University of Michigan Press.

MacKay, K. (2018) 'The ecological crisis is a political crisis', *Millennium Alliance for Humanity*, Available from: https://mahb.stanford.edu/blog/ecological-crisis-political-crisis/ [Accessed 19 July 2023].

Malm, A. (2021) *How to Blow up a Pipeline*, London: Verso.

Marx, K. and Engels, F. ([1848] 2018) *The Communist Manifesto*, Harmondsworth: Penguin/Random House.

Middeldorp, N. and Le Billon, P. (2019) 'Deadly environmental governance: authoritarianism, eco-populism, and the repression of environmental and land defenders', *Annals of the American Association of Geographers*, 109(2): 324–337.

Moskowitz, D. (2023) 'The 10 richest people in the world', *Investopedia*, 3 June, Available from: https://www.investopedia.com/articles/investing/012715/5-richest-people-world.asp#toc-1-elon-musk [Accessed 19 July 2023].

Neocleous, M. (2011) '"A brighter and nicer new life": security as pacification', *Social and Legal Studies*, 20(2): 191–208.

Nixon, R. (2011) *Slow Violence and the Environmentalism of the Poor*, Cambridge, MA: Harvard University Press.

Selwyn, B. (2014) *The Global Development Crisis*, Cambridge: Polity Press.

Simpson, A. (2014) *Mohawk Interruptus: Political Life Across the Borders of Settler States*, Durham, NC: Duke University Press.

Simpson, H. (1973) 'Fascism in South Africa', *African Review*, 3(3): 423–451.

Táíwò, O. (2020) 'Being-in-the-room privilege: elite capture and epistemic deference', *The Philosopher*, 108(4): 61–70.

Thomas, P. (2009) *The Gramscian Moment: Philosophy, Hegemony and Marxism*, Leiden: Brill.

Thunberg, G. (2019) 'The disarming case to act right now on climate change', *TEDxStockholm*, Available from: https://www.ted.com/talks/greta_thunberg_the_disarming_case_to_act_right_now_on_climate_change/transcript [Accessed 19 July 2023].

United Nations Research Institute for Social Development (UNRISD) (2022) 'Crises of inequality: shifting power for a new eco-social contract', Available from: https://cdn.unrisd.org/assets/library/reports/2022/full-report-crises-of-inequality-2022.pdf.

Winlow, S. and Hall, S. (2016) 'Realist criminology and its discontents', *International Journal for Crime, Justice and Social Democracy*, 5(3): 80–94.

Whyte, D. (2014) 'Regimes of permission and state-corporate crime', *State Crime Journal*, 3: 237–246.

INTERSTICE 2

The Gifts of Failure

The Gesturing Towards Decolonial Futures Collective[1]

We chose the word 'gesture' for the name of our collective, Gesturing Towards Decolonial Futures (GTDF), to underscore the fact that decolonization is impossible while our livelihoods continue to be underwritten by colonial violence and unsustainability. The food we eat, the clothes we wear, our health systems and social security, and the technologies that allow us to write about this are all subsidized by expropriation, dispossession, destitution, genocides and ecocides. There is no way around it: we cannot bypass it; the only way is through.

Therefore, although we have a commitment to experiment with decolonial gestures, we are also aware that these will undoubtedly and inevitably fail. Yet, we are also aware that *how* we fail is important. We find that it is actually in the moments when we fail that the deepest learning becomes possible, and that this is usually when we stumble upon something unexpected and extremely useful. Failing generatively requires both intellectual and relational rigour.

Facing failure with accountability, honesty, humility, hyperself-reflexivity, and humour is not easy, but it is a practice that GTDF is trying to develop. In this practice, no one is off the hook; we all have a lot to learn and unlearn, and we will all make mistakes. The first step to moving in this direction is to expand our capacity and disposition to hold space for difficult and painful feedback without feeling overwhelmed, immobilized or wanting to be rescued. In this sense, coddling each other is a way of betraying each other in this process of (un)learning together.

We have created a list of ten hyper-self-reflexivity questions and ten 'potholes' in the road towards decolonization that we have mapped over the years. We use these lists for internal peer reviews of our artistic, pedagogical and cartographic experiments. We share these lists here in recognition that they might be useful for other individuals or groups engaged in decolonial gestures, but without assuming that they would be appropriate or generative in every context.

Hyper-self-reflexivity questions

1. To what extent are you reproducing what you critique?
2. To what extent are you avoiding looking at your own complicities and denials, and at whose expense?
3. What are you doing this for? Who are you accountable to? What is your theory of change? What would you like your work to move in the world?
4. To what extent are you aware of how you are being read by communities of high-intensity struggle? Who (in these communities) would legitimately roll their eyes at what you are doing (finding it indulgent and/or self-infantilizing)?
5. Who/what is this (work) really about? Who is benefiting the most from this work? In what ways could this work be read as self-serving or self-congratulatory?
6. Who is your imagined audience? What do you expect from this audience? What compromises have you had to make in order for your work to be intelligible and relatable to this audience? To what extent can these compromises compromise the work itself? Who are you choosing not to upset and why? How does integrity manifest in your work?
7. To what extent is the politics you are proposing based on the grammar of exceptionalism, entitlements and exaltedness that characterize political engagements within modernity?
8. How wide is the gap between where you think you are at and where you are actually at? Who would be able to help you realize that? Would you be able to listen?
9. To what extent can you respond with humility, honesty, hyperself-reflexivity and humour when your work or self-image is challenged?
10. What would you have to give up or let go of in order to go deeper?

Potholes in the road towards decolonization (for people engaged in low-intensity struggles)

1. Having a critique of colonialism means that you are already decolonized.
   Saying you are doing it does not mean you are actually doing it.
2. Seeing all resistance to authority as anti-colonial.
   Many forms of resistance are inherently colonial and/or imperial.
3. Celebrating all attempts to disrupt colonial patterns as contributing to decolonization.

Most attempts to disrupt colonialism are still grounded in at least some colonial patterns.
4. Extracting, selectively consuming and misinterpreting Indigenous teachings.
   The perceived entitlement to have access to and mastery of Indigenous knowledges is a colonial entitlement.
5. Imagining entanglement as interconnection with beauty only.
   Rather than seeing entanglement as entanglement with 'shit' as well.
6. Emphasizing entitlements and forgetting accountabilities.
   Attempting to transcend privilege without giving anything up.
7. Expecting other people (especially Indigenous people) to shoulder the costs of your learning.
   Attempting to decolonize without considering the impact of your work on Indigenous communities.
8. Confusing self-actualization with decolonization.
   Seeing individual free/creative self-expression as a decolonial gesture.
9. Erasing distinctions between high- and low-intensity struggles.
   Positioning yourself on the basis of individual choice rather than structural location; flattening uneven struggles.
10. Assuming that being a victim of systemic oppression means that you are not complicit in colonialism.
    Although vulnerabilities are unevenly distributed, no one is off the hook. We are all implicated in historical and systemic social and ecological violence.

Note

[1] The copyright to this text will be retained by the Collective.

INTERSTICE 3

Face to Face with the Supercyclone Amphan – Kolkata, 20 May 2020

Sanjana Dutt

Face to face with her

Waking up with a sore throat, I pulled the blanket closer to sleep a little longer. I heard the loud voice of a woman and phrases that included 'Bay of Bengal', 'wind shear', 'storm surges' and 'eye of the cyclone'. These words seemed to come from my father's mobile phone set at a high volume; he regularly follows Facebook videos. Irritated, I finally left my cosy bed and walked towards the balcony. While rubbing my eyes, I saw a flock of crows swirl in the sky and fly somewhere northwest. An elderly couple living opposite our home caught my attention; both of them in their night suits were carrying heavy plant pots from the balcony inside their flat. I turned towards my mother's favourite jasmine plants. She fought with my father to buy 20 of these plants and appeared to love them more than her children. 'Well, weightlifting today', I thought and began relocating these terracotta family members to their safe haven, still sneezing and coughing on my way.

Tropical cyclones have caused 1,942 disasters in the last 50 years, killing 779,324 people and causing US$1,407.6 billion economic losses worldwide (World Meteorological Organization, 2020). As a result of climate change, there has been a marked increase of cyclonic activity in the Indian Ocean in recent years (Mishra, 2014; Sebastian and Behera, 2015). Cyclones in the Bay of Bengal have always been more destructive than other parts of India. The east coast of India witnessed 103 severe cyclones between 1891 and 2000, whereas the west coast experienced 24 storms (NCRMP, 2020). West Bengal, the state where I have lived my entire life, endures these meteorological hazards every alternate spring. Over the last two decades, several severe cyclones have occurred in the Bay of Bengal basin, causing significant damage due to large storm surges (Mondal et al, 2022). But none compared to Amphan, the first supercyclonic storm in the Bay of Bengal since 1999, and one I personally experienced as a resident of Kolkata.

The Regional Specialized Meteorological Centre (RSMC) predicted around a week earlier that Amphan would make landfall between Digha (West Bengal) and Hatiya Islands (Bangladesh) on 20 May. The RSMC predicted maximum sustained winds of 155–165 kilometres per hour, gusting to 185 kilometres per hour and torrential rainfall. Based on this weather forecast, one million people from deltaic Bengal were evacuated.

Somewhere around noon, the sky turned a bluish grey, as if nature decided to use monochrome shades in her palette today. The tall coconut tree around the corner swayed with her long thick leaves as if moaning in pain; the wind had already begun to howl through the leaves. Working from the comfort of my home during the COVID-19 lockdown, I began to record my forthcoming lecture. I called my workplace and requested to reschedule the daily evening class. Call it fate or instinct, but I turned off our Wi-Fi router immediately after the recording. Four seconds later, a peal of thunder nearly jolted our entire house, followed by lightning. Within two hours, low-lying areas were inundated by heavy rainfall. Cyclone Amphan had arrived, with wind speeds of 175 kilometres per hour. Shortly, my television shut down and the entire city was dark for at least six hours.

Winds from Amphan peaked at 165 kilometres per hour ashore at Sagar Island at 2:30 pm, and by 5:00 pm, it had hit Kolkata, located roughly 100 kilometres to the north. The cyclone's 'eye' (centre) measured 30 kilometres in diameter and took an agonising six-and-a-half hours to sweep across the metropolis; its tail eventually left Kolkata at about 11.30 pm as it proceeded towards Bangladesh.

However, the storm did not seem to progress as slowly on the ground. The city and its immediate surroundings were battered by winds ranging from 70 to 90 kilometres per hour, with gusts reaching 112 kilometres per hour in the southern part of Kolkata and 130 kilometres per hour in the northern section of Kolkata. This resulted in at least 10–12 deaths in and around Kolkata.

When scrolling through my friends' Instagram posts, I saw videos of lightning-damaged cable poles and power outages in various parts of the city. Some recorded footage included collapsing trees, rumbling window glass and submerged yellow ambassador taxis. Social media influencers around the world created trending hashtags '#PrayforWestBengal' and '#WeAreInThisTogether'. Some uploaded pictures of rain with romantic Bollywood songs, while others managed a selfie 'with the storm'. Social media sometimes nudges you to romanticize situations when you are less vulnerable. Concerned about my phone's battery life, I cleared all background applications. My mother, sitting in a corner of the room, remained glued to her phone, responding to calls from concerned friends and family members who lived outside Kolkata, asking about our whereabouts and safety. My thoughts flew to a recent Tedx talk about how speaking about your problems actually increases anxiety. The four of us sat in the dark with the occasional light from a candle or phone, as night fell around 6:00 pm. Amid the howling wind, the windows squeaked and water seeped through the walls. My brother, sardonic as always, said 'just act like you are in the sinking Titanic ship right now, only without Jack Dawson around to rescue you'.

The storm broke around 7:20 pm. A shattering sound broke the steadiness of the thundering of rain; a steel panel from the roof of a house flew, as if from nowhere, and collided with our water tank. It felt like one of those scenes from the '2012' sci-fi film. Rainwater from the balcony seeped under the door and into the lounge room. The plants which I tried to protect had already been flipped over by the gusty wind and lay in a mush of clay and sand on the white marble floor. A relative who lived on a twelfth-storey apartment later said that the building seemed to be swaying like it was in an earthquake.

The storm calmed down only after the eye passed over Kolkata, but there were still no signs of electricity. The storm raged until well after 11.30 pm on Wednesday night, with a brief break at about 8:00 pm as its tail escaped the city limits. The city of Kolkata received 244.2 millimetres of rain up until night, and weather officials warned that Amphan's impacts would persist until Thursday evening.

Aftermath

Waking up the next morning was the hardest. It wasn't really a calm after the storm, just an agonizing silence.

Our neighbourhood had a beautiful Earpod Wattle tree, more than 80 years old. The dark, heavily fissured bark lay dead on the sidewalk after crashing through the metallic street railing. A few metres away, a branch from an old deciduous tree had crushed Uncle Roy's red vintage wagon parked on the road.

The daybreak after the cyclone revealed a world unmoored. As I tried to inject some semblance of normalcy into the morning by charging my mobile phone and preparing milk, the microwave oven sparked an electrical short circuit, a minor echo of the previous day's chaos. My father, in a determined flurry, navigated our home's damages, fixing the water tank and painstakingly replacing the windowpanes that had given way to the cyclone's fury. The neighbourhood, too, seemed to wake to a shared, unspoken urgency; phones pressed against ears in a desperate bid to reach the municipality office, the police station and emergency fire services. Their voices, a mix of hope and despair, filled the air, seeking aid, reporting damages, or simply searching for a reassurance that was hard to come by.

The televised footage of three youngsters and a middle-aged homeless mother crying after their temporary shelter was blown away brought me to tears. A video of a little catfish being devoured by a street dog outside the gates of the prestigious Presidency College quickly went viral. The catfish must have been swept away by the floodwaters and was unable to return home. The fish was still alive, and the dog continued to be both intrigued and slightly alarmed by its sharp flapping. The state government faced a formidable challenge in providing safe housing to the homeless amid COVID-19 restrictions. A narrow city lane underlined the overlapping COVID-19 and Cyclone Amphan crises, and the difficult recovery work awaiting officials; as a COVID-19 hotspot and containment zone, its entrance was barred, but it was also submerged in floodwaters.

I called my friends and family; seven of the 12 were still without electricity, and three of their homes had been inundated. The storm completely wrecked the homes of two classmates who lived south of Kolkata city. They posted videos of books floating in the stormwater and food rations washed away. It took ten days for power and water to be restored in the southern districts of Bengal.

Solidarity

Leaves, twigs, branches and shattered tiles were still strewn across the streets of the housing colony long after the rains had receded. Kolkata, the city where I was born, appeared to have slowed to a crawl and was on the verge of dying. Some residents volunteered to help police and security personnel clean major thoroughfares, while others opened their homes to stray dogs and cats.

Some communities rallied to collect donations of food, clothing and medicine from the public, while others went door to door seeking monetary aid. With the help of my neighbours, I collected clothing, grains and legumes for the Leprosy Mission Trust in Kolkata. The Mission provides aid to vulnerable residents of the Sundarbans who had been impacted by Cyclone Amphan. As relatively privileged students in the Geography Faculty of the University of Calcutta, we generated funds to purchase books and other necessities for more vulnerable university students in the state.

The cyclone claimed around 133 lives in India and Bangladesh and many died due to electrocution and house collapse. More than three million people were evacuated before Amphan made landfall; without these prior evacuations, the death toll would have been higher. Several COVID-19 patients perished because of a lack of timely access to oxygen cylinders and expert medical treatment. It took around one month and ten days to clear the uprooted trees and lamp posts on the street; the city recovered slowly.

According to news reports, the estimated economic loss from Cyclone Amphan was around INR 13 billion in India. The personal losses, for sure, were uncountable.

A reporter recounted how a massive banyan tree fell, taking down a small shrine to the Hindu God Shiva, who according to Hindu mythology is regarded as the destroyer of evil. The uprooted tree and temple serve as a reminder that neither gods or prayers could stop the raging Cyclone Amphan.

Amphan etched a formidable portrait in the minds of every human being in West Bengal. The cyclone not only broke down the walls, but also ended up shattering the boundaries of society, making each realize how vulnerable and helpless living beings would be without each other. It made me dwell on the fact that disasters do not discriminate between the rich and the poor, the educated or otherwise; it simply arrives and no matter how strong-armed we are, it could take from us something we fear the most. And in moments like these, it is crucial to stand up for each other, as it could be us the next time. It would not do justice to end describing my experiences and emotions. For me, the following lines from Nobel Laureate and Bengali poet Rabindranath Tagore epitomizes a personal but also collective call for being fearless and compassionate when there is danger and pain. These qualities are necessary in the move towards planetary justice:

> Let me not pray to be sheltered from dangers
> but to be fearless in facing them.
> Let me not beg for the stilling of my pain
> but for the heart to conquer it.
>
> Rabindranath Tagore, 'Fruit Gathering', 1916

References

Mishra, A. (2014) 'Temperature rise and trend of cyclones over the eastern coastal region of India', *Journal of Earth Science and Climatic Change*, 5(9): 227.

Mondal, M., Biswas, A., Haldar, S., Mandal, S., Bhattacharya, S. and Paul, S. (2022) 'Spatio-temporal behaviours of tropical cyclones over the bay of Bengal Basin in last five decades', *Tropical Cyclone Research and Review*, 11(1): 1–15.

NCRMP (2020) 'Cyclones and their impact in India', *National Cyclone Risk Mitigation Project*, Government of India, Available from: https://ncrmp.gov.in/cyclones-their-impact-in-india/ [Accessed 30 June 2020].

Sebastian, M., and Behera, M.R. (2015) 'Impact of SST on tropical cyclones in North Indian Ocean', *Procedia Engineering*, 116: 1072–1077.

World Meteorological Organization (2020) 'Tropical cyclones', Available from: https://public.wmo.int/en/our-mandate/focus-areas/natural-hazards-and-disaster-risk-reduction/tropical-cyclones [Accessed 6 May 2020].

PART III

Learning and Living with Climate Change as Situated Solidarity

9

Planetary Justice and Decolonizing Pedagogy: Teaching and Learning in Solidarity with Country

Aleryk Fricker

Always was, always will be

For millennia, human communities, cultures and societies have thrived, living sustainably by caring for Country to ensure the ongoing success of consecutive generations. As a species, we have migrated and populated every continent in the world with the exception of Antarctica. Our ancestors survived through ice ages, sea-level changes and significant seismic events, and still we have endured.

First Nations cultures are the oldest continuous cultures in the world and have existed on the Australian continent and adjacent islands for over 60,000 years (Malaspinas et al, 2016). During this time, given that culture is not innate, these many cultures were taught to subsequent generations of children and adults through complex and specific pedagogical processes developed to support their growth, wellbeing and learning outcomes. One such pedagogical focus was learning on, with and through Country (Moran et al, 2018). On the Australian continent and adjacent islands, my ancestors lived their lives ensuring that there was not only enough to be shared in their present, but that there would also be enough to be shared in my present time – that is, until 235 years ago, when the British began their wholesale invasion and colonization of Australia and disrupted the ways of my Dja Dja Wurrung ancestors.

The declaration of the Australian continent and adjacent islands as *terra nullius* supported the dispossession of Country from the First Nations peoples and, in turn, the disruption of the Country-based pedagogical practices that had supported learning for millennia. To further disrupt the learning of First Nations children, the European schooling system present in England at the time was then imposed upon the First Nations peoples where, like all Eurocentric pedagogical approaches, learning on, with and through Country was absent.

This chapter explores the current global climate emergency, concepts of Country (see Chapter 2) and the means of decolonizing pedagogy. I argue that a recognition of First Nations pedagogical sovereignty, as a challenge to colonial pedagogical approaches, created in solidarity with Country and First Nations stakeholders, will support efforts toward planetary justice.

My position in this research

This chapter has been composed on the lands of the Wurundjeri people of the Eastern Kulin Nations and I pay my respects to the Traditional Owners and Elders past and present. I write this as a proud and sovereign Dja Dja Wurrung man who was born and has lived all my life off Country, with little connection to my ancestral lands. Despite my distance from my Country, I recognize the need to care for the Country that I live on and ensure that I am showing my respect for the laws and lore of the Traditional Owners of where I live and work.

Unprecedented events

Australia, like the rest of the world, is facing a climate crisis unlike any other that has been experienced by any generation of the human species (Archer and Rahmstorf, 2010). We are facing multiple climate catastrophes, including global warming and climate change (Bandh, 2022), biodiversity loss (Wood et al, 2000) and ocean acidification (Raisman and Murphy, 2013), as well as many other crises related to the unsustainable living conditions present across the globe. Despite some of the gains that national governments have enacted, this is far below what needs to be achieved to delay or reverse the current climate trajectory (IPCC, 2022).

Beyond the global impacts of climate change, Australia has also experienced a significant drying of the southern parts of the continent (CSIRO, 2018), and this has increased the frequency of devastating bushfires (Lindenmayer and Taylor, 2020) and has impacted the native plants and animals that can be successfully cultivated, as my ancestors did, at these latitudes (Pascoe, 2014).

These crises will dominate the current and many future generations to come. It is imperative that the education available to students reflects a solutions-based approach that will empower them to navigate through, act to limit, and reverse the impacts of the unsustainable living conditions that have been created and spread globally through European colonization.

Understanding Country

A central part of the understanding of why and how we had to care for Country was gifted to us through interactions with our Elders, families, wider community, ancestors, spirits, and Country itself. It was predicated on a profound understanding that we were in relationship with Country, which supported the question of *why* Country had to be cared for, and the millennia of scientific practices and knowledges that supported the question of *how* to care for Country. During these times, we understood what it was to live and learn sustainably in solidarity with Country, and the impacts this would have on us and the generations to come.

Before exploring the topic of learning in solidarity with Country, the concept of Country should be better understood. Primarily, it should also be noted that there is no one definition of Country that is accepted by all First Nations groups across the Australian continent and the adjacent islands; there is at least one concept of Country for each cultural grouping. Nonetheless, there are some general concepts that relate to Country that are consistently accepted as being part of many different Dreamings.[1]

A starting point to explore what Country is to define what it is not. For First Nations Peoples, Country is not a commodity or something to be exploited. Country is not something to be owned or something to be used to stratify society, as has been applied in many Western contexts (Van de Mieroop, 2004). For First Nations Peoples, Country is an area of land that is directly related to a group of First Nations people, and the relationship with Country extends beyond time and is recorded through stories laid down in Country (Southern Cross University, 2019). Country is also alive and intelligent, and provides the people with everything they need (Southern Cross University, 2019). Country is a place where, prior to colonization, people would travel over and sustain themselves, and where the people rested and where the children played (Jackson-Barrett and Lee-Hammond, 2018).

In another part of the continent of Australia, Country is defined as part of a First Nations ontology where animals, rocks, winds, tides, emotions, spirits, songs, and humans speak (Bawaka Country et al, 2015). This is where they all have language and law, and where they all send messages and communicate with each other (Bawaka Country et al, 2015). The concept of Country is also closely linked to a vibrant and sentient understanding of space and place bounded through interconnectivity, and where Country and everything it encompasses is an active participant in the world, both shaping and creating it (Bawaka Country et al, 2015). Beyond the physical structures of Country, as a concept it also exists within the people as a model for being human in a proper way (Southern Cross University, 2019).

In short, Country is us, our food, medicine and resources. Country is our lore and law, and our responsibility. Country is our stories, languages, dances, songs, sacred places and ceremonies. Country is our cultures, our spirits, ancestors, our past, present and future. County is our libraries, universities, schools, classrooms and, importantly, our teacher. People need Country as Country needs people, as when we care for Country, Country cares for us.

Allies, accomplices, converts, comrades

Prior to considering how we can be in solidarity with Country, it is crucial to ensure that the nomenclature used to name the stakeholders and the process is fit for purpose. The language and the labels we use when naming and defining the work that needs to be done to address systems of oppression are important. To name a group of people or a specific process has the power to establish paradigms of meaning that impact how these are interpreted, experienced and applied in the real world. These paradigms also contain the 'operative assumptions about how we see and comprehend the basic nature of a problem' and, as such, it 'necessarily informs the kind of solutions or responses we identify' (Harvey, 2014, pp 14–15). This means that, whether we are conscious of it or not, 'language shapes our imaginations in ways that bear deeply on the possibilities of action' (Jantzen, 2020, p 275). Therefore, it is a crucial step to get the language right to influence and encourage the best possible outcome for Country.

The term ally is problematic. In many of my interactions with non-Indigenous people, usually the moment they describe themselves as an ally in a First Nations context, a red flag is immediately apparent. This has also been explored in research focused on how there is often little understanding of allyship, and this often leads to performative allyship that is at best not helpful and at worst actively harmful for the marginalized group it relates to (Kalina, 2020). In this context, the language is important. The most common conception of allies or allyship is often applied to a context when two nation states are negotiating treaties relating to military actions.

In the Australian context, one of the most important treaties establishing allyship is the Australia, New Zealand, United States Security Treaty (ANZUS Treaty) signed in 1951. As part of this alliance and treaty, all three nation states – Australia, Aotearoa and the US – had military obligations for mutual support, in the event that one or more members of the alliance were attacked (Robb and Gill, 2015). Where this provides an apt reflection of the challenges of allyship on a personal level is when Aotearoa left the formal alliance in the mid-1980s over a dispute with the US, relating to declaring its country and surrounding region as a non-nuclear area.

This meant that Aotearoa could no longer guarantee access to its ports for US nuclear-powered naval assets and therefore could not be part of the treaty. The consequence of this decision reflects the weakness of allyship on a personal level. There was a period of suppressed diplomatic communication and repercussions, but several decades later, Aotearoa is still on excellent terms with the US. This means that one of the weaknesses of allyship on a personal level is that those purporting to be allies have the power to leave when the circumstances become too difficult with limited or no repercussions. They get to continue to enjoy their lives and privilege, while those who are marginalized continue to fight against the systems of oppression.

This has left many feeling (and rightly so) that allyship is 'superficial, self-serving, and shallow' and has allowed white allies to 'associate oneself with fashionable ideas or movements or people, while absolving oneself of the responsibility for serious personal transformation, committed action, and shared risk taking' (Jantzen, 2020, p 274). This was highlighted in activist literature in a Turtle Island context, where it was succinctly put that: 'Everyone calls themselves an ally, until it's time to do some real ally shit' (Xhopakelxhit, 2015).

The obvious limitations of the term 'ally' mean that language has begun to change in some contexts and the term 'accomplice' is gaining some prominence. Accomplice was originally coined in grassroots activist contexts and has become more common in academic and mainstream discourse (Jantzen, 2020). The key contrast between these two terms is that rather than relying on the largesse of the privileged ally, an accomplice is a person who is associated with others in crime or wrongdoing (Jantzen, 2020). As such, this language positions the actions of the privileged accomplices as seeking to 'puncture the safe and satisfied detachment of allyship, by emphasising the risk, action, and personal transformation' (Jantzen, 2020, p 274) required to be successful in this space.

The term 'accomplice', despite being an improvement on the term 'ally', is also not without its limitations. To use an analogy again to identify the limitations, I use the plot from just about every 'bank heist'-style film that has ever been produced. In these films, a group of people come together as accomplices, planning and executing a robbery in often very dramatic circumstances. As part of this process, it is made clear that all accomplices either have the same risk and reward, or a proportional reward based on their risk in the operation. The issue with this is that when it comes to dismantling the oppressive systems and structures, First Nations accomplices still have the burden of greater risk. Furthermore, as Tuck and Yang (2012) state, decolonization is not a metaphor and, as such, there will ultimately need to be a redistribution of land and wealth to offset the colonial dispossession of First Nations people. To borrow from the analogy given earlier, this would be the risk burdening one accomplice, while the rewards went to another, and this uneven distribution would not work.

A third proposed term is 'convert' (Jantzen, 2020). This term has its origins in the works of James Cone (2008) from an African-American context and has been developed further through the work of Harvey (2012), Lloyd (2016) and McGee (2017). A convert is described as a person who has undergone some process of repentance that results in a change in that person's whole thinking – this is a conversion (Cone, 2008). From the point of conversion, a convert is someone who has gone through a total and radical transformation of one's entire self, where their actions place their life at stake for a new way of being in the world (Cone, 2008). Cone (1997, p 221) also argues that a true experience of conversion involves a person being 'reborn anew in order to struggle *against* white oppression and *for* the liberation of the oppressed'.

Despite the power that this language and approach could have in the context of building solidarity between Country, First Nations people and non-Indigenous Australians, this term has some limitations. Primarily, the act of conversion is problematic in the Australian context. Various religious groups were active as part of the genocide in Australia, establishing many missions and reserves across the continent and adjacent islands, with the express purpose of converting and civilizing the 'natives' (Maddison, 2014). Furthermore, the association with a religious concept is also problematic, as it implies that all the work done in the space to challenge oppression specifically relates to the experiences that will occur after death. The issue is that this is an ontological incommensurability with many First Nations concepts of Dreaming and cycles of life and death, as well as conceptions of the living and nonliving worlds (Hume, 2004).

Overall, the current terminology that relates to non-Indigenous people acting in ways to actively dismantle neocolonization in Australia are not fit for purpose and create contexts where the labels 'ally', 'accomplice' and 'convert' are co-opted and changed to render them ineffective. It could be argued that this set of terms are part of the 'settler moves to innocence' used to relieve settler anxiety and proceed as usual, as discussed by Tuck and Yang (2012). In addition, these are all human-centred concepts that fail to account for the importance of Country as part of the nonhuman world. As such, there is a need to re-examine the language relating to how non-Indigenous people can take on the necessary labour to dismantle the structures and ideas that privilege them in all ways.

The appropriate term proposed in this chapter is 'comrade'. When considering the ongoing colonization of Australia and the implications for First Nations people, the only non-Indigenous people who can be effective with us are comrades. Comradeship is defined as a form of political relation among those who desire collectivity and see themselves on the same side of a struggle for communism (Lessing, 1962). Comradeship has also been defined as promising liberation from 'the constraints of racist patriarchal capitalism' and gaining a 'new relation born of collective political work toward an emancipatory egalitarian future' (Dean, 2020, p 155).

The description of comrade also complements the work of solidarity that can be achieved through action. Solidarity is described as 'a set of practices directed toward specific political goals rooted in the specificities of relationships' (Gaztambide-Fernández et al, 2022, p 253). Furthermore, solidarity involves relationships, intentions and ethical commitments (Gaztambide-Fernández and Matute, 2014). No one can be in solidarity alone and, as such, it requires 'deliberate attention to particular relationships and to the dynamic entanglements that produce the similarities and differences that animate these relationships' (Gaztambide-Fernández et al, 2022, p 254). The final reason why solidarity is supported by comradeship is that, like comradeship that seeks to collectively liberate the oppressed, solidarity too seeks transformation grounded in reciprocity and consent (Gaztambide-Fernández et al, 2022).

Furthermore, comradeship also recognizes the reality of the hegemonic power structures that also dominate and restrict those in the privileged sections of society. As argued by Freire (1970, p 31), liberation is a painful process that can only occur through the humanization of all people and when the oppressed enlist 'in the struggle to free themselves'. Freire also argues that as the consciousness builds in the minds of the oppressors, this does not necessarily result in establishing solidarity with the oppressed; rather, it requires agency on behalf of the oppressor to 'enter into the situation of those with whom one is solidary' and that this is 'a radical posture' (1970, p 31). He also states that 'true solidarity with the oppressed means fighting at their side to transform the objective reality which has made them these "beings for another"' (1970, p 31).

As such, in order to be a comrade and in solidarity with First Nations people and contexts, one must fight at our sides and 'show up to meetings we would miss, do political work we would avoid, and try to live up to our responsibilities to each other' (Dean, 2020, p 155) – in other words, be prepared to pay a personal cost in time, wealth or risk that needs to be paid as part of the fight against oppression and recognize that it is the collective effort and cost paid by many that build success in challenging oppressive power structures. We need comrades at our sides, amplifying our voices, leveraging their power and privilege, taking our direction and leadership, and collectively speaking back to power both within and beyond the classroom. Only through this unwavering solidarity will we be able to make meaningful change.

Decolonizing pedagogy

From the discussion of comradeship and being in solidarity with First Nations people and Country, the focus must shift to the role of education in both preparing the next generation to endure and respond to the climate emergency, and act in ways that support planetary justice.

In the Australian context, our current education system is a colonial construction that was implemented as part of the ongoing colonization of the Australian continent and the adjacent islands (Austin, 1972). As part of this construction, pedagogical approaches mirroring those in Europe were implemented, designed to shape the learning in the classroom that would produce productive workers, perpetuate the hegemonic power of middle- and upper-class white society, and resemble conditions in a European Industrial Revolution workplace (Mitch, 1999).

As part of constructing schooling and education in these ways, the pedagogical approaches used in these classrooms prioritized specific rote ways of learning and a 'one-size-fits-all' approach. Beyond this limited engagement with the science of teaching and learning, these pedagogical approaches were often limited to domains that in no way engaged with the environment beyond the classroom. This lack of engagement created a distance between the students and their environment, and generated a distorted binary that school was the place one went to 'learn' and everywhere else was not a place for learning (Gaudelli, 2020). By excluding the environment beyond the classroom for learning, Western pedagogies effectively excluded any possibility to incorporate learning through engaging with Country.

This historical binary relation means that any opportunities for teachers and students to teach and learn in solidarity with Country requires that pedagogy be decolonized and, through this process, challenge how pedagogy is still constructed, experienced and practised as part of a neocolonial Australian schooling system. Decolonizing pedagogy is an emerging research focus in Australia and has been described as a process of decolonizing environments to foster relational teaching and learning (Shahjahan et al, 2022). Beyond this general definition, decolonizing pedagogy focuses on foregrounding the collaborative nature of knowledge production in the classroom, the institution and with external communities (de Carvelo et al, 2016; Parker et al, 2017; Padilla, 2019).

Further scholarship has explored how the process of decolonizing pedagogy can manifest in two main ways. The first is through challenging the 'banking model' of teaching from teacher to student in a one-way process, and instead focusing on the co-production of knowledge between teacher and students by pedagogies that incorporate active learning and critical reflexivity (Shahjahan et al, 2022). The second is a focus on the integration of cultural and spiritual pedagogical practices, including Indigenous practices, affective, embodied learning, rituals and ceremony (Adefarakan, 2018; Ng, 2018; Wong 2018).

In the Australian context, both these approaches are relevant in the pedagogical research and innovation that is manifesting in schools, teacher training qualifications and pedagogical practices in the classroom. There has been a resurgence in First Nations pedagogical practice that has directly contributed to the push to decolonize pedagogy in the classroom. Some notable examples of these pedagogies in Australia are Eight Ways based on eight approaches to teaching and learning from Wiradjuri contexts (Yunkaporta, 2009), Two-Way Learning which focuses on including First Nations approaches to learning as part of the Australian education system (Purdie et al, 2011), and On Country Learning focusing on what Country can directly teach us (Bawaka Country et al 2015; Jackson-Barrett and Lee-Hammond, 2018; Moran et al, 2018). Ultimately, teaching and learning in solidarity with Country will only occur through a wider project of decolonizing pedagogy and the associated understanding that we all have much to learn from, on and in solidarity with Country. This approach will also extend the understanding of, and support for, the implementation of other First Nations pedagogical approaches.

Teaching in solidarity with Country for planetary justice

There are several significant adjustments to key educational contexts that will have to be made to teach in solidarity with Country. These relate to curriculum, educational places and spaces, and pedagogy and the ways in which learning is conceptualized and constructed in Australia. The first aspect requires critical self-reflection on the neocolonial ontologies and epistemologies that have constructed Country as 'land', which is both inert and commodified by the whole education workforce. From this point, Country can then be conceptualized in different ways and, in a non-Indigenous context, go beyond the 'humanization' of people theorized by Freire (1970) to consider, for want of a better term, the 'humanization' of Country.

This humanization therefore allows for the largely non-Indigenous education workforce to recognize that we have, and have always had, a relationship with our environment. This relationship in the Australian continent up until the colonization began was reciprocal, with obligations and responsibilities to care for Country, which in turn would care for the people. This would create the consciousness within the education workforce to 'enter into the situation of those with whom one is solidary' (Freire, 1970, p 31) – that is, enter the situation with Country and then show up to meetings, do political work and try to live up to our responsibilities to each other, despite not wanting to (Dean, 2020). This would ultimately allow for Country to be emancipated with us as part of our collective egalitarian future (Dean, 2020) – just like it was in Australia before colonization.

Curriculum

The political work of teaching and learning in solidarity with Country in terms of curriculum will require the curriculum architects and the teaching staff who implement it to reframe the Cross Curriculum Priority of sustainability (ACARA, 2022). This priority is described as focusing on exploring the 'knowledge, skills, values and world views necessary for people to act in ways that contribute to a sustainable future' (ACARA, 2022). It has four subtopics that consist of Systems, World Views, Design and Futures, and these four subtopics allow for students to engage with the concept of sustainability in a range of contexts. One of the challenges is that sustainability is positioned as a specific aim and purpose of engaging with Country. However, this positioning does not align with a First Nations conception of sustainability which is an *outcome* of caring for Country and not a focus in and of itself (Fricker et al, in press).

In addition, the positioning of sustainability as a Cross Curriculum Priority means that rather than sustainability being centred across every aspect of a school embedded in caring for Country, it is positioned in a precarious place in the curriculum and is at risk of being ignored (Salter and Maxwell, 2015). To teach and learn in solidarity with Country will require a reform to the curriculum where Country is embedded at every possible point, as well as making all curriculum decisions based on how they will directly and indirectly impact Country.

Place and space

Teaching and learning in solidarity with Country also requires a revision and adjustment of the places and spaces where teaching and learning occur. The colonial project of establishing an education system modelled on the European version designed to create productive citizens ignored Country-based pedagogies, and so Country was removed from the classroom. This has created generations of classrooms almost entirely divorced from Country in highly regulated spaces, which often include standardized design and close control (Woolner et al, 2022).

In this context, to teach and learn in solidarity with Country is to include, and make visible, Country in the teaching and learning spaces within the school, and to extend the learning beyond the school grounds. Before colonization, Country was our first teacher. From the earliest ages, First Nations children would be taught by Country, sometimes curated by Elders or other community members, but also often just between Country and the learner through observations, insights and inspiration directly from Country. Country always knew that if it could teach us, it could continue to sustain us. By including and making Country visible in the places we teach and learn, we allow Country to again teach us the *how* and *why* we should care for Country and give us the chance to fight for planetary justice in all its forms.

In the Australian context, planetary justice will be achieved in solidarity with Country; reflecting that planetary justice in a First Nations context refers to fighting for socioecological justice for humans and nonhumans with an emphasis on multiscalar aspects of justice for earth systems globally (Dryzek and Pickering, 2019). Central to this fight for planetary justice is the concept of healing as a form of justice for both people and Country (Youngblood Henderson and McCaslin, 2005). In the Australian context, and in line with the decolonial process, focusing on the distribution of land and wealth provides an opportunity to heal the people through the alleviation of struggling under a neocolonial capitalist system, and heal Country by allowing access and implementation of First Nations land management practices. Furthermore, by beginning to dismantle the neocolonial capitalist systems, we directly challenge the extractive economic systems that are driving the global climate emergency.

In a practical sense, Australian schools have a great opportunity to consider how they could begin this process through the incorporation of the various 'kitchen garden' programmes with native plants included. Another could be schools being active in local land rehabilitation programmes. These would provide opportunities for students to engage directly and begin to build their relationship with Country. Furthermore, schools supporting students to engage as political citizens with their parliamentary representatives through literacy or civics and citizenship activities would provide opportunities to advocate on behalf of Country to achieve planetary justice.

Finally, teaching and learning in solidarity with Country would also entail the expansion of on-Country learning programmes. By building and expanding these programmes, students can then learn with both Country and the local Traditional Owner community. In doing so, they not only build their relationship with Country, but also enhance their learning either through Elder curation of, or directly from, Country. By doing so, students can experience learning as it has been practised for millennia across the continent of Australia and the adjacent islands – through building relationships and getting to know Country. It is only once we have a relationship with and get to know Country that we can fight for Country to achieve planetary justice.

Conclusion

Country and the First Nations Peoples do not need any more allies; they have only ever gotten us so far. We do not need any accomplices expecting their reward, and we certainly do not need any converts. We need comrades, those who will do what is right despite the personal cost – those who are willing to show up and do the required work. Country has always been our first teacher, our life giver and our classroom, and is central to our Dreaming. In the context of global environmental disaster, there has never been a more crucial time to act in solidarity with Country and fight for planetary justice. Teaching and learning in solidarity with Country is crucial in order to build authentic relationships with Country that can teach us that to care for Country is to care for ourselves and each other, and that our future survival will be dependent on our ability to fight for planetary justice in solidarity with Country.

Note

[1] The term 'Dreaming' is a mistranslation by some of the early European linguists in Australia. This term does not translate well into English, but it is the ontological, epistemological, methodological and axiological viewpoint of a First Nations person as well as the stories of creation. This concept is predicated on the relationship between all things including the living and nonliving realms and temporal considerations, and informs every decision that a First Nations person makes. Each First Nations culture has a different Dreaming and this is often shaped by Country and the community that live there.

References

Adefarakan, T. (2018) 'Integrating body, mind, and spirit throughout the Yoruba Ori: critical contributions to a decolonizing pedagogy', in S. Batacharya and Y. Wong (eds) *Sharing Breath: Embodied Learning and Decolonization*, Edmonton: Athabasca University Press, pp 229–252.

Archer, D. and Rahmstorf, S. (2010) *The Climate Crisis: An Introductory Guide to Climate Change*, Cambridge: Cambridge University Press.

Austin, A. (1972) *Australian Education, 1788–1900: Church, State and Public Education in Colonial Australia*, Carlton: Pitman Pacific Books.

Australian Curriculum and Assessment Authority (ACARA) (2022) 'Understand this cross-curriculum priority: sustainability', Available from: https://v9.australiancurriculum.edu.au/teacher-resources/understand-this-cross-curriculum-priority/sustainability [Accessed 16 March 2023].

Bandh, S. (2022) *Climate Change: The Social and Scientific Construct*, Cham: Springer.

Bawaka Country, Wright, S., Suchet-Pearson, S., Lloyd, K., Burarrwanga, L., Ganambarr, R., Ganambarr-Stubbs, M., Ganabarr, B. and Maymuru, D. (2015) 'Working with and learning from Country: decentring human author-ity', *Cultural Geographies*, 22(2): 269–283.

Cone, J. (1997) *God of the Oppressed*, Maryknoll: Orbis Books.

Cone, J. (2008) *Black Theology and Black Power*, Maryknoll: Orbis Books.

CSIRO (2018) 'Australia's weather and climate continues to change in response to warming global climate', Available from: https://www.csiro.au/en/research/environmental-impacts/climate-change/state-of-the-climate/previous/state-of-the-climate-2018/australias-changing-climate [Accessed 4 March 2023].

De Carvalho, J., Cohen, L., Corrêa, A., Chada, S. and Nakayama, P. (2016) 'The meeting of knowledges as a contribution to ethnomusicology and music education', *World of Music*, 5(1): 111–133.

Dean, J. (2020) 'Capitalism is the end of the world', *Meditations*, 33(1–2): 149–158.
Dryzek, J. and Pickering, J. (2019) *Planetary Justice: The Politics of the Anthropocene*, Oxford: Oxford University Press.
Freire, P. (1970) *Pedagogy of the Oppressed*, London: Penguin.
Fricker, A., Cooper, G., Kilmartin-Linch, S. and Sheffield, R. (in press) 'Promoting First Nations understandings of sustainability in both teacher professional development and undergraduate course learning', in M. Rosano (ed) *The Routledge Handbook of Global Sustainability Education and Thinking for the 21st Century*, New York: Taylor & Francis.
Gaudelli, W. (2020) 'The trouble of Western education', *On Education. Journal for Research and Debate*, 3(7): 1–4.
Gaztambide-Fernández, R., Brant, J. and Desai, C. (2022) 'Toward a pedagogy of solidarity', *Curriculum Inquiry*, 52(3): 251–265.
Gaztambide-Fernández, R. and Matute, A. (2014) 'Pushing against', in J. Burdick, J. Sandlin, and M. O'Malley (eds) *Problematizing Public Pedagogy*, New York: Routledge, pp 52–64.
Harvey, J. (2012) 'What would Zacchaeus do? The case for disidentifying with Jesus', in G. Yancy (ed) *Christology and Whiteness: What Would Jesus Do?* New York: Routledge, pp 84–100.
Harvey, J. (2014) *Dear White Christians: For Those Still Longing for Racial Reconciliation*, Cambridge: William B. Eerdmans Publishing Company.
Hume, L. (2004) 'Accessing the eternal: dreaming "the Dreaming" and ceremonial performance', *Zygon*, 39(1): 237–258.
Intergovernmental Panel on Climate Change (IPCC) (2022) *Climate Change 2022: Impacts, Adaptation and Vulnerability*, Contribution of Working Group II to the Sixth Assessment Report of the Intergovernmental Panel on Climate Change, Cambridge: Cambridge University Press.
Jackson-Barrett, E. and Lee-Hammond, L. (2018) 'Strengthening identities and involvement of Aboriginal children through learning on Country', *Australian Journal of Teacher Education*, 43(6): 86–104.
Jantzen, M. (2020) 'Neither ally, nor accomplice', *Journal of the Society of Christian Ethics*, 40(2): 273–290.
Kalina, P. (2020) 'Performative allyship', *Technium Social Sciences Journal*, 11: 478–481.
Lessing, D. (1962) *The Golden Notebook*. New York: Simon & Schuster.
Lindenmayer, D. and Taylor, C. (2020) 'New special analysis of Australian wildfires highlight the need for new fire, resource, and conservation policies', *Proceedings of the National Academy of Sciences*, 117(22): 12481–12485.
Lloyd, V. (2016) 'For what are whites to hope?', *Political Theology*, 17(2): 168–181.

Maddison, S. (2014) 'Missionary genocide: moral illegitimacy and the churches in Australia', in J. Havea (ed) *Indigenous Australia and the Unfinished Business of Theology: Postcolonialism and Religions*, New York: Palgrave Macmillan, pp 31–46.

Malaspinas, A. et al (2016) 'A genomic history of Aboriginal Australia', *Nature*, 538(7624): 207–214.

McGee, T. (2017) 'Against (white) redemption: James Cone and the Christological disruption of racial discourse and white solidarity', *Political Theology*, 18(7): 542–559.

Mitch, D. (1999) *The British Industrial Revolution: An Economic Perspective*, New York: Routledge.

Moran C., Harrington, G. and Sheehan, N. (2018) 'On Country learning', *Journal of the Design Studies Forum*, 10(1): 71–79.

Ng, R. (2018) 'Decolonizing teaching and learning through embodied learning', in S. Batacharya and Y. Wong (eds) *Sharing Breath Embodied Learning and Decolonization*, Edmonton: Athabasca University Press, pp 33–54.

Padilla, N. (2019) 'Decolonizing Indigenous education: an Indigenous pluriversity within a university in Cauca, Colombia', *Social and Cultural Geography*, 22(4): 523–544.

Parker, P., Smith, S. and Dennison, J. (2017) 'Decolonising the classroom', *Tijdschrift voor Genderstudies*, 20(3): 233–247.

Pascoe, B. (2014) *Dark Emu: Aboriginal Australia and the Birth of Agriculture*, Broome: Magabala Books.

Purdie, N., Millgate, G. and Bell, H. (2011) *Two Way Teaching and Learning: Toward Culturally Reflective and Relevant Education*, Camberwell: ACER Press.

Raisman, S. and Murphy, D. (2013) *Ocean Acidification: Elements and Considerations*, Hauppauge: Novinka.

Robb, T. and Gill, D. (2015) 'The ANZUS Treaty during the Cold War: a reinterpretation of US diplomacy in the Pacific Southwest', *Journal of Cold War Studies*, 17(4): 109–157.

Salter, P. and Maxwell, J. (2015) 'The inherent vulnerability of the Australian curriculum's cross-curriculum priorities', *Critical Studies in Education*, 56(2): 1–17.

Shahjahan, R., Estera, A., Surla, K. and Edwards, K. (2022) '"Decolonizing" curriculum and pedagogy: a comparative review across disciplines and global higher education contexts', *Review of Educational Research*, 92(1): 73–113.

Southern Cross University (2019) *Gnibi Wandarahn: Innovate Reconciliation Action Plan, March 2019–March 2021*, Lismore: Southern Cross University.

Tuck, E. and Yang, K.W. (2012) 'Decolonisation is not a metaphor', *Decolonization: Indigeneity, Education and Society*, 1(1): 1–40.

Van de Mieroop, M. (2004) *The Ancient Mesopotamian City*, Oxford: Oxford University Press.

Wong, Y.R. (2018) '"Please call me by my true names": a decolonizing pedagogy of mindfulness and interbeing in critical social work education', in S. Batacharya and Y. Wong (eds) *Sharing Breath: Embodied Learning and Decolonization*, Edmonton: Athabasca University Press, pp 253–277.

Wood, A., Stedman-Edwards, P. and Mang, J. (2000) *The Root Causes of Biodiversity Loss*, London: Earthscan.

Woolner, P., Thomas, U. and Charteris, J. (2022) 'The risks of standardised school building design: beyond aligning the parts of a learning environment', *European Educational Research Journal*, 21(4): 627–644.

Xhopakelxhit (2015) *Everyone Calls Themselves an Ally, Until It's Time to Do Some Real Ally Shit*, Portland: Microcosm Publishing.

Youngblood Handerson, J. and McCaslin, W. (2005) 'Exploring justice as healing', in W. McCaslin (ed) *Justice as Healing: Indigenous Ways*, St. Paul: Living Justice Press, pp 3–12.

Yunkaporta, T. (2009) 'Aboriginal pedagogies at the cultural interface', PhD thesis, James Cook University.

10

Towards Transformative Social Resilience: Charting a Path with Climate-Vulnerable Communities in the Indian Sundarbans

Jenia Mukherjee, Amrita Sen, Kuntala Lahiri-Dutt and Aditya Ghosh

Introduction

Can 'social resilience' be realized through community-level and co-produced plans designed within the immediate context of climate-risk-prone lifeworlds of coastal and deltaic communities? Drawing on participatory action research in villages situated within the Indian Sundarban Biosphere Reserve (SBR),[1] this chapter argues that these stakeholders not only anticipate climate disasters, but that they also have the capacity to examine existing and complex social challenges. This capacity helps communities to address both vulnerabilities and coping mechanisms more closely, taking into consideration multiple anxieties and disruptions that occur over long periods of climate crisis. To this end, this chapter describes how we engaged with local community members, as well as consulted with a range of other actors who associate themselves with policies and action plans related to the mitigation of climate risk. The productive engagement demonstrates that in order for social resilience to be transformed into long-term bounce-back strategies, cross-disciplinary perspectives must be employed, and resilience must not be confined to a concept within its usual 'ecological' niche.

The chapter concludes that there is a strong need to deploy co-produced policies and strategies towards risk mitigation and the building of local-level social resilience as an effective strategy towards planetary justice. Better prospects are achieved through convergences, collective recoveries and co-produced adaptation knowledge. This perspective builds on the empirical exploration of the Sundarbans mangrove forest, an archipelago of river islands that accommodates the largest stretch of littoral mangroves in the world. Our focus here is on the Indian part of the SBR, a unique space that is shared by humans and nonhumans, among them the Royal Bengal Tiger (*Panthera tigris tigris*) and fish-eating crocodiles.

The SBR has been a focus of international forums on climate change. Divergent interest groups – from mainstream sustainability scientists to bureaucrats and nongovernmental organizations – have highlighted the precarious situation of this dynamic region under a changing climate. Policy responses have essentially focused on interventionist conservation measures: of tigers and other fauna, and the unique land and waters that support the area's rich biodiversity. Other efforts include the expansion of ecotourism, better mangrove plantation design, the construction of higher embankments made of concrete, and the building of cyclone centres. Yet, together these efforts have produced a vision of an apocalyptic future in which scientific conservation discourses have usurped the voice and agency of the poor who continue to live, as the local saying goes, with tigers on land and crocodiles in water (*dangay bagh, jale kumir*).

Against this background, this chapter presents local stories of the Anthropocene by bringing to the fore everyday lived realities; struggles and community adjustment 'tactics' to combat volatile ecologies. These stories, in response to climate and planetary injustice, challenge rigidly crafted boundaries, such as Indigenous versus modern, state versus local, and vulnerability versus resilience, by factoring in 'temporality' and fluid understandings of the community. We argue that resilience that attunes to human and more-than-human socioecological vulnerabilities is informed by community-based adaptation to change, where local and Indigenous knowledge systems play a critical role.

These local stories were elicited through interactive workshops and focus group discussions, case studies, transect walks, informal onsite conversations, participant observations and participatory mapping. These ethnographic and participatory tools of research were deployed in what are generally considered to be some of the remotest villages located on three islands: Kumirmari, Frasergunj and Tipligheri. The stories show that in the Anthropocene, community-level action plans can be formulated to build resilience if they are co-produced through knowledge, skills and adaptation practices. These knowledges have the potential to heal the ruptures of abrupt socioenvironmental transformations and place-based anthropocentric shocks.

Setting aside the introduction and conclusion, the chapter is comprised of four sections. First, we discuss how we plan to reframe the social resilience framework. Then we introduce the field site in the disaster-prone SBR, which has the potential to frame new understandings of social resilience. The third section discusses the conceptual feedback loops in deploying social resilience through empirical research. We conclude with a short commentary on the ethics of scaling up networked social resilience through co-produced adaptation knowledge that moves towards planetary justice.

Problem statement: reframing social resilience

Over the last few decades, global warming has severely impacted the delicate ecological balance of the SBR. The people of the Sundarbans frequently encounter tropical cyclones and their spiralling impacts, including breaching of embankments, saline water intrusion, waterlogging of agricultural land and so on. Yet despite a significant body of research proposing technical and physical fixes for sustained capacity-building arrangements, there is still a need to explore how postdisaster crises can be tackled using social resilience built from existing community skills. The disaster-induced anxieties and the co-produced knowledge that evolved in solidarity with the communities in this landscape move beyond 'unviable futures' (Paprocki, 2022, p 2) and offer lessons for building new ways of thinking about social resilience strategies.

In a 2020 edition of *Nature*, Bai et al (2020) express a strong and urgent need to share plans and build solidarity networks among a range of different actors to tackle the impacts of future anthropocentric risks. They suggest that the development of 'networked functional resilience', through sharing and co-evolving disaster risk plans and actions, can be useful in thinking about and planning for planetary justice (Bai et al, 2020, p 520). These networked solidarities are particularly important in coastal areas like the SBR that are affected by climate hazards. Drawing on Biermann and Kalfagianni (2020, p 2) and alongside an understanding of what lived experience of justice means for diverse communities, we understand planetary justice to be a transition from 'a normative debate on planetary justice towards an empirical debate on what conceptualizations of justice different actors in global environmental politics actually support'. Introducing networked solidarities into the domain of planetary justice can expand the idea of (and the need for) synergies for effective climate risk abatement. We explore this idea qualitatively by 'integrating multiple stakeholder interactions and exchanges that can converge' (Bhattacharya et al, 2023, p 5), as well as by identifying core capacities of social resilience and knowledge co-production through situated adaptive practices (SAPs).

The concept of 'resilience' has remained heavily anchored in bioecological sciences (Walker, 1993), and often remains 'inadequate', masking power relations (Cannon and Müller-Mahn, 2010, p 623), depoliticizing social structures and unconsciously reinforcing the status quo by overlooking those mechanisms that put people at risk in the first place (Pelling and Manuel-Navarrete, 2011). Most existing frameworks and approaches pay little attention to human agency and empowerment, politics and power relations, ideologies, risk perception and the diversity of cultural values, or capacities for human (rather than environmental) transformation (Bahadur and Tanner, 2014, p 23). From a socioecological perspective, all these narratives are valid because they form elements of latent and dominant feedback loops that require articulation for a nuanced understanding of vulnerability-reducing and resilience-building responses in a collaborative framework (Maru et al, 2014, p 337). In this chapter, we argue that resilience building must include pragmatic ways of understanding local lifeworlds more closely than before, with an explicit recognition of not only environmental challenges but also growing social complexities and fragilities.

In providing a detailed analysis of knowledge co-production, Norström et al (2020) facilitate our understanding of the significance of a cluster of participatory and transdisciplinary research. Our empirical framework and methodological expertise helped us to execute project deliverables in such a way that we were able to consciously listen to community concerns and issues, such as that of adequate skills yet inadequate hands-on deployment knowledge. This led to meetings between fishery scientists and communities, who directly exchanged ideas about scaling up existing skills and practices. Following Norström et al (2020), who discuss two broad approaches – normative and descriptive – we feel that this is how sustainability research can gain traction.

While the normative (and hence pragmatic) approach investigates co-production as a deliberate collaboration between different actors and networks to achieve common goals, the descriptive approach provides weight to the reciprocal relationship between science and society in which knowledge is continually shaped and co-produced within the existing social order. However, the two approaches are intertwined; participation and validation of pluriversal possibilities occur alongside science–society interactions. In his trailblazing work *Pluriversal Politics: The Real and the Possible*, the Colombian anthropologist Arturo Escobar unravels new perspectives on alternative realities, inviting researchers to open their minds (ontologically) to pluriversal possibilities: 'a world where many worlds fit' (2020, p 9).

Drawing on transdisciplinary and participatory research, Norström et al (2020, p 183) identify four principles of knowledge production to encourage academia, civil society and user groups 'to engage in meaningful co-productive practices and address the sustainability challenges of the Anthropocene'. These principles are context-based, pluralistic, goal-oriented and interactive. They not only identify and suggest convergence among various top–down policy-driven schemes and missions, but also upscale bottom-up needs-driven skills, expertise and grassroots (community) practices prevalent in place-based contexts. Such convergence helps to effectively examine practical risk scenarios in specific vulnerable landscapes, like the SBR, leading to the formulation of an integrated climate risk adaptation framework, co-produced through interinstitutional and intersectoral initiatives. Our case study is a stimulating addition to research on social resilience, knowledge co-production and their intersections for gradually yet 'competently' dealing with disasters. But how do we build on Norström et al's (2020) work?

The strength of the social resilience approach is that it fosters possibilities for hope amid climate vulnerability and risk. Knowledge co-production in local places emerges through a series of exchanges, involvement and commitment between multiple stakeholders (Mukherjee et al, 2022). Such social resilience, which moves beyond state–community binaries and mainstream tenets of the 'adaptation' discourse, elicits the possibility of scaling up situated adaptive practices.

Various allied disciplines, such as climate studies and environmental humanities that aim to understand the intersections between socioecological systems and disaster management, have explored and employed the concept of resilience. The numerous disaster events that have occurred around the world in recent decades have significant social impacts due to their increasingly devastating effects on local communities (Saja et al, 2019). Yet, the application of a social resilience framework to disasters is inconsistent and confusing without a comprehensive understanding informed by examining a multitude of divergent spatial contexts.

In this chapter, we contest the claim that effective resilience-based capacity building and disaster management is the responsibility of formal governance, external to multiple lived realities. Indeed, Keck and Sakdapolrak (2013, p 5) argue that a combination of three factors – 'coping capacities' (an actor's capacity to deal with and overcome various adversities), 'adaptive capacities' (ways to draw lessons from the past and adapt to upcoming obstacles in daily life) and 'transformative capacities' (creation of systems of institutions that promote wellbeing and resilience in the face of future crises) – can appropriately be leveraged for an institutionalization of social resilience.

Several questions therefore converge around our work on social resilience. First, how should we, as humans, act during and after crises? Second, how do we deal with and act in volatile ecologies (Mukherjee et al, 2023) thickly inhabited by ecosystem-dependent communities? Third, how do we integrate the knowledge base, skills and expertise of plural actors across a long temporal range? Finally, how can the reframing of social resilience contribute to strategies that deal with chronic disasters and reduce aggregate vulnerabilities, inequities and power dynamics inherent in natural processes?

Study area: the Indian SBR

The SBR, a United Nations Educational, Scientific and Cultural Organization (UNESCO) World Heritage Site, is recognized for its physical attributes, including its plentiful wildlife, biodiversity and aquatic resources. In total, it covers an area of 9,630 square kilometres, 4,263 square kilometres of which are uninhabited forest. The remaining 5,367 square kilometres are a densely populated transition zone. The region is not a continuous landmass; rivers traverse the landscape and divide it into separate islands. Some of the forests are protected and are known as the Sundarban Tiger Reserve (STR). The region is subject to tidal currents that are created by river distributaries as they move towards the Bay of Bengal.

The SBR is a fragile and vulnerable ecosystem that has experienced pronounced effects of climate change, and as a result has received a great deal of academic attention. Rising sea levels, accompanied by stronger tidal waves, have inundated and eroded chunks of land, depleting mangroves and leading to the extinction of wildlife species. According to a projection by the World Wildlife Fund, the sea-facing erosion-prone islands have lost between 3 per cent and 32 per cent of land, and this is primarily attributable to cyclonic storms (World Wildlife India, 2023). A large number of government reports emphasize that rising sea levels will render much of the land uninhabitable in the coming years.

The SBR is inhabited by economically marginalized and socially vulnerable communities. But global attention focuses on the area's prized biodiversity and wildlife rather than on the existing threats to human lives. The narratives of the people, and their everyday struggles in the ever-eroding volatile landscape, remain invisible and disregarded (Ghosh, 2018; see also Chapter 4). The social aspects of uncertainty and risk show that 'climatic uncertainties intersect[ing] with other socio-economic drivers of change also can end up creating new uncertainties and vulnerabilities especially for poor and powerless people constraining their livelihood choices' (Mehta et al, 2019, p 1534). Shaped by extreme natural events, the landscape of the Sundarbans is physically vulnerable, and it is made more so by haphazard and disconcerted governance and risk planning (Ghosh et al, 2022). Dominant approaches to ecological conservation and the design of coping mechanisms are embedded in institutional arrangements.

However, the SBR is not a homogeneous ecological or social landscape. The southern blocks are more susceptible to the hazards and associated risks of climate change, owing to their remote location, socioeconomic marginalization and long-term government apathy regarding infrastructural interventions. In contrast, the extreme northern limits of the SBR are populated by well-knit settlements in close proximity to the city of Kolkata. The land surface of these northern settlements is stable because they are geographically distant from forests and rivers, and therefore from the threats of erosion.

The survival and livelihood of the population that inhabits the island corridors of the forest in the southern limit of SBR is the focus of the present study. Forests and rivers are the two main sources of livelihood in these villages, but the settlements are dissected by many rivulets and are thus located on a relatively unstable land surface. Jalais (2004, p 17) refers to these islands as being 'on the move'. Here the tidal action of the river channels is more active as the rivers reach their final journey before joining the sea. The fragile nature of the landscape, which is formed by the frequently changing course of the river channels, abruptly erodes land surfaces, and thus supports few lucrative economic options (Jalais, 2004; Sen, 2022).

The 'riskscapes' of Gosaba: rethinking social resilience

The southernmost block of Gosaba, in the South 24 Parganas district of West Bengal, accommodates the village *panchayat* where we conducted our fieldwork. It is the epitome of 'disruptive risks' and awaiting 'disruptive resilience' (Bahadur and Dodman, 2020, p 1). The severely poverty-stricken island villages of Gosaba are repeatedly exposed to cyclones, floods and tiger attacks, which are the result of both natural and human-induced factors, including the vicissitudes of a faulty conservationist ethic as well as risk governance. In 2020, Gosaba was one of the worst-affected blocks in the SBR (Ghosh, 2020) when the villages of Satjelia, Lahiripur and Kumirmari were severely devastated by Cyclone Amphan. Cyclone Amphan led to a greater dependence on forests due to agrarian disruptions caused by increased soil salinity. This was compounded by in-migration during the nationwide COVID-19 lockdown. Widespread loss of employment across the country because of the suspension of economic activities and associated large-scale reverse migration further exacerbated the impacts of the cyclone, which included the destruction of homesteads and traditional livelihoods, such as agriculture and fishing. These multipronged stresses have underscored weaknesses in existing disaster management systems.

Gosaba's additional and contextual vulnerability is shaped by its territorial and topographical attributes. This easternmost block, with an average elevation of less than four metres above sea level, also sits beside the STR. Thus, in their pursuit of forest fringe fishing, villagers are subject to frequent attacks by the Royal Bengal Tiger, which is known for its interisland migration patterns and ecoadaptive behaviours to dynamic intertidal habitat transformations. Moreover, regulations governing access to forests have played a primary role in the deprivation, dispossession and exploitation of village communities. The buffer zones within the STR, unlike the core zone within the STR, are legally accessible for the purposes of ecotourism. Forest fishers (who pursue fishing activities in the forest creeks and estuaries) enjoy some rights and are permitted to catch fish if they possess boat licences issued by the government. Yet, just over two thirds of these licences are operational, and the licensing system is manipulated and monopolized by rich agriculturalists and intermediaries, locally known as *aratdaars* or *khotidaars*, whose ancestors were engaged in forest fishing (Sen and Patnaik, 2017). The inhabitants of Gosaba 'reside with risks' (Pathak, 2023), their multiple stressors entangled and shaping the fate of the island villages through the complex production of the realities of risk (Müller-Mahn et al, 2018).

In the SBR in general and in the more isolated blocks of Gosaba in particular, existing technophysical solutions remain superficial; they are prescriptive designs, external to the context, and largely ignore the existing knowledge base of the communities encountering and living with risks. We argue that beyond the 'predict and provide' approach (Adger, 2000), the nurturing of social resilience through co-produced knowledge can have a sustained impact on community wellbeing through the design and implementation of locally appropriate solutions to risks. Within the Gosaba block, we selected Kumirmari *gram panchayat* as our study area. Our initiative in Kumirmari is a microcosmic tapestry involving grand plans and aspirations to thrive in the Anthropocene. In January 2021, we conducted in-depth interviews, focus group discussions and surveys. We chose Kumirmari as a selective case study owing to its growing exposure to climate hazards and instances of community adaptation. Communities in Kumirmari are primarily composed of immigrants from Bangladesh, with a sizeable number of tribal, Christian, Muslim and other caste populations.

Situated knowledge co-production in Kumirmari

Our understanding of the correlation between knowledge and action as a feedback loop, constantly operating and transforming on a spatiotemporal plane, is circumstantial. Our aim is to conduct ethical research that does justice to endangered lives and livelihoods in this coastal area.

Our field visit to the SBR followed months of lockdown during the COVID-19 pandemic. We were thus motivated to optimize every opportunity and to learn as much as we could from the residents through ethnographic research. More than ever, we were empathetic to the village communities who we knew had suffered much more than us during the lockdowns and cyclones, which had turned rural frontiers and vulnerable villages upside down.

The local grassroots organization Sundarban Jana Sramajibi Mancha (SJSM) agreed to guide our research in Gosaba. Village representatives received us at the *ghat* on the bank of the river and accompanied us on our village walks. They also familiarized us with the island villages and acted as mediators during meetings and discussions, ensuring 'focused' conversations between us and the local communities. Some of these onsite meetings were also attended by *panchayat pradhans* and other dignitaries from the *panchayat* committees.

The villagers were critical of influential political leaders who they believed had appropriated, monopolized and hijacked discussions. Moreover, they felt that these individuals were only interested in promoting their political agendas by channelling infrastructure funding and relief distribution during extreme events, instead of revealing numerous structural bottlenecks. We were able to further assess the situation by walking through the villages, and listening to groups of women who conveyed their multilevel stress and trauma. They told us about the frequent cyclones, repeated tiger attacks and exploitation by forest officials, who were ever ready to fine and tyrannize forest fishers, pushing them towards deprivation and penury. 'We are afraid of the forest officials more than the tigers' was a statement made by a tribal forest fisher from Tipligheri during one of the focus group discussions conducted at a marketplace at dusk.

One gram panchayat, *one village, unique attributes*

On day three, our boat reached Kumirmari village island (Figure 10.1). This island seemed different from the others we had visited; there were signs of material prosperity and community aspirations, hope and trust, despite the constant threat of impending crises. The island is inhabited by 4,344 households, of which 3,000 own land and ponds that are mainly used for inland fisheries as the locals reported. Kumirmari has 26 locally excavated *khals* (canals) and commercial inland fishing is practised in

Figure 10.1: Map of Kumirmari

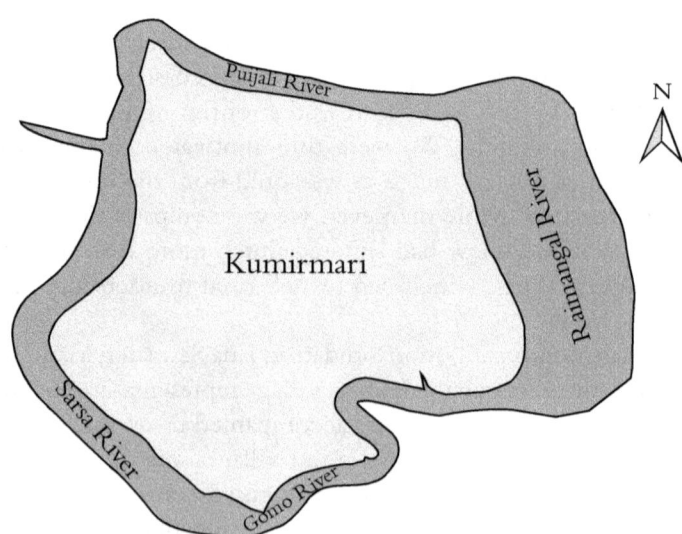

Source: Developed by Raktima Ghosh

some of these. The canals are cooperatively managed, and the profits are shared equally between the villagers who lease the canals for seasonal fish production, and social events such as festivals, which are organized by local clubs.

During interviews other village community members were somewhat passive in relation to climate and associated risks, but the villagers of Kumirmari were enthusiastic in sharing their coping mechanisms. They had been exposed to risks and thus abatement strategies for a long time. With limited options for forest fishing and agrarian experiments, inland fishing provides people with an alternative livelihood. Moreover, with the increased risk of tiger attacks and stringent rules regarding mangrove-based fish and crab harvesting, fishing in domestic and cooperative ponds and cooperative canals in Kumirmari has recently emerged as a more lucrative and viable option. Yet, myriad issues such as fish disease, inappropriate fish seeds and prolonged salinization of land in the postcyclone period, continue to disrupt the wellbeing of fishers. They voiced their concerns about their inadequate knowledge of how to continuously adapt and adjust during postcyclone periods, with gushing waters submerging ponds and saline water remaining stagnant, impacting on and across various stages of inland fish production.

The villagers therefore asserted the importance of gaining scientific expertise through targeted training by the Department of Fisheries, the government of West Bengal and other technical organizations in order to successfully conduct pisciculture in ponds and ensure returns on investment. The villagers complained about the sporadic and infrequent efforts of the Department of Environment in providing training programmes. They also noted that when programmes are offered, they are organized in places that are difficult to reach. Moreover, expertise and knowledge are seldom disseminated at the village level, meaning that fishers are not empowered or given the confidence to pursue inland fishing activities as a profitable venture. Despite these challenges, the concerted efforts of fishers, the availability of ponds for the majority of households, a willingness to adopt inland fishing as the major source of income and the desire to be trained in adaptive fishing practices marked out Kumirmari as distinct from the other island village communities. This points to the possibility that Kumirmari villagers can move from being active coping actors to adaptive managers through locally planned interventions and appropriate training from experts via action research across sustained phases.

The axiological departure: lessons learnt

Our field research in Kumirmari was motivated by the occurrence of Cyclone Yaas in May 2021. As well as the knowledge to action (K2A) small grants

Table 10.1: Multiple stages of K2A project execution

Stage	Component
Analysis	Situational analysis and explorative mapping of stressors and responses
Design	Design of prototype outputs with information from stage one
Development	Development of project outputs
Implementation	Dissemination (in pilot mode) in targeted platforms
Evaluation	Feedback and review; finalization of outputs – larger dissemination

Source: Mukherjee (2022)

opportunity for projects with the potential to translate academic research into tools and processes that support awareness, advocacy and transformation of the UN Sustainable Development Goals (SDGs) under the aegis of Swissnex. Our aim was to design, develop and disseminate a training module documenting the situated adaptive practices shaping inland fishing through knowledge co-production with multiple stakeholders, including the SJSM, our academic research project team, government trainers and experts, local government institutions and the fishers themselves. The project was designed according to the analysis-design-development-implementation-evaluation model (ADDIE) (Table 10.1), which is aligned to the SDGs that address sustenance (SDG 1, 2 and 8), the social wellbeing of fishers (SDG 3, 5 and 10), local knowledge and innovation (SDG 9), and relies on inclusive and collaborative partnerships (SDG 17).

Knowledge co-production through multimodal methodology

Our project followed multiple methodological formats to co-produce information from an improvised and contextually meaningful version of 'participatory systems mapping' (Barbook-Johnson and Penn, 2021). This included trust-building exercises between academia, practitioners and user groups to onsite workshops forging face-to-face exchanges among stakeholders. We also facilitated a nuanced 'strengths, weaknesses, opportunities and threats' (SWOT) analysis (Table 10.2) to map the challenges and opportunities of inland fishing experimentation in Kumirmari to identify more sustainable outcomes.

The prototype training module used extensive infographics and was produced in both English and vernacular languages. It drew on insights from the fishers themselves, secondary literature, including reports from the Food and Agricultural Organization and the United Nations, and meetings between the project team and scientists from the Central Inland Fisheries Research Institute (ICAR). However, during the later stages of the fieldwork,

Table 10.2: Summary of the SWOT analysis

Strengths	Plural nature of fishing practice, including traditional knowledge of inland fishing
	Availability of land and water for fishing
	Self-taught expertise in hatching
	Involvement of women
Weaknesses	Lack of necessary guidelines relating to systemic cultivation
	Poor quality of locally available fishing
	Fish diseases and infections
	Access to market
	Lack of cold storage facilities
	Inadequate technical knowhow
	Exploitation by intermediaries
Opportunities	Profit generation in commercial inland cultivation encouraging landless or small plot holders to come up with cooperatives
	Unwillingness among the younger generation to venture into the forest
	Enthusiasm and involvement of women
	Traditional knowledge traits and tenurial security
Threats	Climatic instability
	Limited options for landless or the households without a water body
	Loss of indigenous fish species and stocks

Source: Participatory multistakeholders' workshop, Kolkata, January 2022

the villagers and the SJSM insisted that without practical action, the training module, in the form of booklets distributed among households and posters in the *panchayat* office, would have a negligible impact.

During the workshops, fishers emphasized the need to discuss issues directly with government trainers to learn about specific-measurable-attainable-relevant-timebound (SMART) implementation strategies, so that their coping mechanisms relating to inland fishery could advance into transformative capacities. This would enable them to become socially resilient; to respond to and prepare for an array of immediate physiosocial risks and disaster events in the near future. Innovative practices would arise from appropriate training, relying heavily on locally contingent scenarios such as cost-effective infrastructures, intergenerational knowledge and the skillsets of fishers.

Going beyond the purview of project outputs and enacting outcomes as part of the organic and evolutionary process in this participatory action research, the project team conducted a place-based, needs-based onsite workshop during which villagers were able to communicate directly with the principal scientist of the Central Inland Fisheries Research Institute.

This allowed the project team to map the areas, degrees and methods of intervention in successful inland fishery. The research team also facilitated knowledge co-production and documented the entire process to test the processual trajectory and future viability of the resilience-based action plans. The team administered a questionnaire asking villagers about micro, meso, and macro issues, including technical expertise, cultural traditions, ecological parameters, and social and institutional governance scenarios shaping inland fisheries in Kumirmari. However, discussions during the workshop were not restricted to fishing activities, but also covered other important aspects such as market links, possibilities for cold storage facilities and allied activities such as mushroom farming, crab fattening and handicrafts to optimize the pisciculture initiative. Thus, the onsite workshop gave the villagers (and the research team) an understanding of the multiple agencies that could be approached.

Our Indian Institute of Technology Kharagpur research team, with the support of all stakeholders as full partners, are now geared up to test the prototype participatory training manual in identified ponds and cooperative canals in 2023. In addition, the process will be rigorously documented, planned and fostered through the participation, involvement and engagement of the co-produced knowledge of the collaborative epistemic community.

Social resilience, knowledge co-production and situated adaptive practices

In this chapter we have formulated a methodological approach to deconstruct social resilience as a dominant buzzword across multiple fields of study and practice. In doing so, we have attempted to test viabilities emerging from multiple communities towards knowledge co-production. Using pragmatic methodological tools like SWOT, ADDIE and SAPs, we applied social resilience as an outcome of the adaptive capacities of multiple communities to tackle disasters through deliberate planning strategies and implementation, relying on their skills and suggesting pathways to scale these up. These methodological tools better enable people in vulnerable ecological systems 'to respond and recover from disasters' (Cutter et al, 2008, p 599). Learnings from the past and the anticipatory capacities of social actors remain embedded in key components that allow the system to absorb shocks and impacts and deal with hazards and posthazard scenarios, facilitating the capacity of the social system to reorganize, transform and learn, and appropriately respond to threats. Hence, the impact remains deep-rooted and sustained through preparatory concerted actions, and constant experimentation and improvization of 'tactics'.

Our focus on 'training practices' disseminates existing knowledge on adaptation to the local level, enabling the place itself to act 'as a system of social adaptation' during climate crisis scenarios rather than focusing abruptly and inconsistently on external adaptation action plans devised by the state (Lyon, 2014, p 1009). As the empirical observations show, many of our efforts were shaped by the participation of multi-actor institutions and networks that remain enablers for social communities to optimize existing resource bases, learn from past experiences and devise constructive ways of dealing with common challenges. Additionally, we showed how even catastrophes can lead to opportunities, making way for experimentation, innovation and development. Obrist et al (2010, p 291) argue that social resilience has the potential to enhance 'capacities of individuals, groups and organizations to deal with threats more competently'.

Conclusion

Employing a co-produced conceptual thematic of social resilience and testing its empirical credibility through rigorous analysis of field exercises has enabled us to learn about how people actually adapt to change instead of telling them how they should adapt. Knowledge of local practices, in addition to the ADDIE and SWOT models, has shown that a networked and synergistic multistakeholder approach can be the most critical driver in mobilizing collective efficacy during hazardous climatic conditions. In Kumirmari we uncovered opportunities for a fuller development of community capabilities that can generate a 'core set of social resilience indicators' (Kwok et al, 2016, p 197). Our field workshops, community-capacitation programmes and training modules, for example, are innovative in developing skills that can contribute to disaster risk reduction. Inland aquaculture, previously the least-viable livelihood option owing to lack of capacity building, is now practised widely on the island. It has proved to be a livelihood alternative that has lowered forest dependence and risks to life from tiger attacks.

With leading personnel from the Fisheries and Forest Department assisting the fishers in this process of capacity-building, our SWOT analysis demonstrates pathways to deployment of plural and traditional knowledge in fishing alongside governance-based technical expertise. In addition to bolstering livelihood opportunities and skills, we provided hands-on training to the community to deploy the acquired skills for comprehensive 'social wellbeing' during a range of potential risk scenarios. In the translation of this conceptual framing into practical action, we implemented the SMART strategies, which revealed avenues for averting risks during future scenarios. The uniqueness of our proposed framework is its relevance to local realities – fragility, insecurity, conflict, and informality – that are amplified in disaster scenarios. These place-based vulnerabilities are always unique, and we show that many of these vulnerabilities represent major lessons for a comprehensive development of social resilience derived from a robust risk management plan.

Our contribution in this chapter illuminates a possible approach for moving towards planetary justice. It provides ideas for alternatives, engagement and opportunities for mapping how intersectional solidarities and viabilities can materialize between multiple stakeholders through the practical involvement of academic researchers in the field. It transcends claims regarding the role of resilience, instead framing resilience as central to both social justice and sustainable ecosystems. Such intersections have proved to be not only viable but pragmatic alternatives in thinking about planetary justice. Such a vision of planetary justice lies in the potential of successful policy convergences, collective postdisaster recoveries, co-produced adaptation knowledge, scaling up situated practices, and the mutual synthesis of 'local' and 'translocal' understandings of hazard mitigation. This transition, and its scope in policy and practice, highlights prospects for solidarity networks that can be leveraged to achieve comprehensive social resilience in the face of future environmental challenges. The differential and participatory capacities of social actors are key components that can be used as 'bounce-back' strategies, facilitating an essentially social system in responding to future climatic threats.

Note

[1] We thank all respondents for the time they gave to the empirical part of this research. This research was funded by the EU-ICSSR-sponsored EqUIP Project (2019–2022) 'Towards a Fluid Governance: Hydrosocial Analysis of Flood Paradigms and Management Practices in Rhone and Ganges Basins (India, France and Switzerland)', the Swiss sponsored CLOC K2A initiative, the SSHRC-funded 'Vulnerability to Viability' project and the ICSSR-funded 'An innovative adaptation and social resilience framework: using a multi-sectoral approach to address climate vulnerability issues in the Indian Sundarbans' major project (2022–2024). We would like to acknowledge EqUIP, CLOC K2A and SOR4D ENGAGE grants in actuating theoretical and practical aspects pertaining to this research on social resilience in the Indian Sundarbans.

References

Adger, W.N. (2000) 'Social and ecological resilience: are they related?', *Progress in Human Geography*, 24(3): 347–364.

Bahadur, A. and Dodman, D. (2020) *Disruptive Resilience: An Agenda for the New Normal in Cities of the Global South*, London: International Institute of Environment and Development.

Bahadur, A. and Tanner, T. (2014) 'Transformational resilience thinking: putting people, power and politics at the heart of urban climate resilience', *Environment and Urbanization*, 26(1): 200–214.

Bai, X., Nagendra, H., Shi, P. and Liu, H. (2020) 'Cities: build networks and share plans to emerge stronger from COVID-19', *Nature*, 584: 517–520.

Barbrook-Johnson, P. and Penn, A. (2021) 'Participatory systems mapping for complex energy policy evaluation', *Evaluation*, 27(1): 57–79.

Bhattacharya, S., Mukherjee, J., Choudry, A. and Ghosh, R. (2023) 'Reifying "River": unpacking pluriversal possibilities in rejuvenation surrounding the Adi-Ganga of Kolkata', *Frontiers in Water*, 5: 1070644.

Biermann, F. and Kalfagianni, A. (2020) 'Planetary justice: a research framework', *Earth System Governance*, 6: 100049.

Cannon, T. and Müller-Mahn, D. (2010) 'Vulnerability, resilience and development discourses in context of climate change', *Natural Hazards*, 55(3): 621–635.

Cutter, S.L., Barnes, L., Berry, M., Burton, C., Evans, E., Tate, E. and Webb, J. (2008) 'A place-based model for understanding community resilience to natural disasters', *Global Environmental Change*, 18(4), 598–606.

Escobar, A. (2020) *Pluriversal Politics: The Real and the Possible*, Durham, NC: Duke University Press.

Ghosh, A. (2018) *Sustainability Conflicts in Coastal India*, Gewerbestrasse: Springer Nature.

Ghosh, H. (2020) '"Our Sundarban is unrecognisable": life after Cyclone Amphan wrecked the island', *The Wire* [online], 28 May, Available from: https://thewire.in/environment/sundarbans-cycle-amphan-ground-report [Accessed 10 August 2022].

Ghosh, R., Mukherjee, J., Pathak, S., Choudry, A., Bhattacharya, S., Sen, A. and Patnaik. P. (2022) 'A situational analysis of small-scale fisheries in the Sundarbans, India: from vulnerability to viability', V2V Working Paper 2022-4, V2V Global Partnership, University of Waterloo, Canada.

Jalais, A. (2004) 'People and tigers: an anthropological study of the Sundarbans of West Bengal, India', PhD dissertation, London School of Economics and Political Science.

Keck, M. and Sakdapolrak, P. (2013) 'What is social resilience? Lessons learned and ways forward', *Erdkunde*, 67(1): 5–19.

Kwok, A.H., Doyle, E.E., Becker, J., Johnston, D. and Paton, D. (2016) 'What is "social resilience"? Perspectives of disaster researchers, emergency management practitioners, and policymakers in New Zealand', *International Journal of Disaster Risk Reduction*, 19: 197–211.

Lyon, C. (2014) 'Place systems and social resilience: a framework for understanding place in social adaptation, resilience, and transformation', *Society and Natural Resources*, 27(10): 1009–1023.

Maru, Y.T., Smith, M.S., Sparrow, A., Pinho, P.F. and Dube, O.P. (2014) 'A linked vulnerability and resilience framework for adaptation pathways in remote disadvantaged communities', *Global Environmental Change*, 28: 337–350.

Mehta, L., Srivastava, S., Adam, H.N., Bose, S., Ghosh, U. and Kumar, V.V. (2019) 'Climate change and uncertainty from "above" and "below": perspectives from India', *Regional Environmental Change*, 19(6): 1533–1547.

Mukherjee, J., Lahiri-Dutt, K. and Ghosh, R. (2023) 'Beyond (un)stable: chars as dynamic destabilisers of problematic binaries', *Social Anthropology/Anthropologie Sociale*, DOI: https://doi.org/10.3167/saas.2023.04132305.

Mukherjee, J., Bhattacharya, S., Ghosh, R., Pathak, S. and Choudry, A. (2022) 'Environment, society and sustainability: the transdisciplinary exigency for a desirable Anthropocene', in M.I. Hassan, S. Sen Roy, U. Chatterjee, S. Chakraborty and U. Singh (eds) *Social Morphology, Human Welfare and Sustainability*, Cham: Springer Nature, pp 35–64.

Müller-Mahn, D., Everts, J. and Stephan, C. (2018) 'Riskscapes revisited: exploring the relationship between risk, space and practice', *Erdkunde*, 72(3): 197–214.

Norström, A.V., Cvitanovic, C., Löf, M.F., West, S., Wyborn, C., Balvanera, P., Bednarek, A.T., Bennett, E.M., Biggs, R., de Bremond, A. and Campbell, B.M. (2020) 'Principles for knowledge co-production in sustainability research', *Nature Sustainability*, 3(3): 182–190.

Obrist, B., Pfeiffer, C. and Henley, R. (2010) 'Multi-layered social resilience: a new approach in mitigation research', *Progress in Development Studies*, 10(4): 283–293.

Paprocki, K. (2022) 'On viability: climate change and the science of possible futures', *Global Environmental Change*, 73: 102487.

Pathak, S. (2023) 'Residing with risks: everyday story of a woman in dried fish practice', in E. Thrift et al (eds) *Dried Fish Matters: Exploring the Social Economy of Dried Fish*, St Johns, NL, Canada: TBTI Global Book Series, n.p.

Pelling, M. and Manuel-Navarrete, D. (2011) 'From resilience to transformation: the adaptive cycle in two Mexican urban centers', *Ecology and Society*, 16(2): 11.

Saja, A.M.A., Goonetilleke, A., Teo, M. and Ziyath, A.M. (2019) 'A critical review of social resilience assessment frameworks in disaster management', *International Journal of Disaster Risk Reduction*, DOI: 10.1016/j.ijdrr.2019.101096.

Sen, A. (2022) *A Political Ecology of Forest Conservation in India: Communities, Wildlife and the State*, Abingdon: Routledge.

Sen, A. and Pattanaik, S. (2017) 'Community-based natural resource management in the Sundarbans: Implications of customary rights, law and practices', *Economic and Political Weekly*, 52(29): 93–104.

Walker, B.H. (1993) 'Rangeland ecology: understanding and managing change', *Ambio*, 22: 80–87.

World Wildlife India (2023) 'Conservation challenges', Available from: https://www.wwfindia.org/about_wwf/critical_regions/sundarbans3/conservation_challenges_in_the_sundarbans/ [Accessed 25 October 2019].

11

Profane Knowledge, Climate Anxiety and the Politics of Education

Callum McGregor, Beth Christie and Marlies Kustatscher

Introduction

This chapter critically reflects on our involvement in a collaborative project on Education for Climate Justice (ECJ) held in Scotland in 2021. Through bringing our own reflections into dialogue, we develop an analysis of the politics of education by critically considering the relationship between the institutionalized disavowal of profane knowledge and climate anxiety. By 'profane knowledge', we mean knowledge deemed by powerholders to be too radical or ideologically charged. In Scotland, young people have been key actors in demands for climate justice and the prioritization of the climate crisis in education. Taking these demands seriously, we brought together diverse participants over a series of three events to examine what it means to truly place social justice at the heart of efforts to address the climate crisis in education. The three events were held in February, March and April 2021, and each event accommodated between 50 and 80 participants.[1] By diverse stakeholders, we mean young people, climate activists, community-based adult learners, teachers, youth workers, academics, trade unionists and policy workers. While our analysis emerges from this particular context and partial perspective, we hope that it resonates beyond its origins given the increasingly widespread recognition that education cannot adequately respond to the immediate global sustainability challenges we face without addressing social injustices.

We situate our dialogue within the context of the politics of sustainability education by critically considering its positioning in the broader ideological struggles that shape education policy and practice. Our argument unfolds around the ambivalent discourse of climate anxiety, as we reflect critically on the ways in which diverse young people's emotion work is tied to a contradictory situation whereby the existential threat of climate emergency is mainstreamed, yet educational engagement with radical ideas is either tacitly or coercively prohibited.[2] Marlies begins the dialogue by locating the ambivalent discourse of climate anxiety in a wider discussion of the emotional politics of education for climate justice. Callum continues the dialogue by reflecting on young people's need for self-directed learning and considers whether climate anxiety experienced by young people might be understood as a symptom of contradictions and disavowals in educational policy and practice. Finally, Beth builds on these reflections to consider issues of teacher agency more directly. To conclude, we suggest what education for the climate crisis might look like if social justice was taken seriously. Throughout, we offer our reflections from various standpoints including teacher perspectives, young people's experiences and our own reflections as researchers working in this space.

The emotional politics of education for climate justice: interrogating discourses of climate anxiety (Marlies)

Discourses of climate emergency and justice are highly emotive; popular and academic discussions have increasingly paid attention to these emotional aspects. For the purposes of this chapter, I view emotions not as internal states of individuals, but rather as social and cultural practices which contribute to how issues, identities and groups are constructed (Ahmed, 2014). It follows that emotional dimensions should not be seen as separate from what are sometimes perceived as the purely factual elements around the climate crisis. Instead, 'the possibilities for more productive forms of discussion and action on climate change rest with more complete understandings of the diversity of social experiences of this phenomenon, including its affective content' (Jones and Davison, 2021, p 191).

For ECJ, this means that there is a need to integrate emotional dimensions with the facts, politics and activisms of climate justice. There are multiple ways in which attention to emotions can benefit ECJ, from exploring their role as motivators or inhibitors of action to supporting young people's wellbeing in the face of undeniable adversity. While such approaches are sometimes couched under concepts like emotional literacy, there are limits to such an instrumental view of emotions, which sees them as states that eventually need to be channelled in 'correct' ways for productive outcomes, and as individualized rather than collective. Indeed, school-based approaches to fostering emotional literacy have been described as forms of self-government informed by neuroscientific technologies of producing ideal citizens (Gagen, 2015).

An increasing field of research has focused on emotions and the climate crisis, and particularly on how emotions shape people's actions. Jones and Davison (2021) broadly identify these studies as exploring grief, anxiety, apathy and hope, as drivers or inhibitors of social engagement against the climate crisis. Some of these studies are particularly sensitive to the ambivalent nature of emotions. For example, Norgaard (2011) suggests that apathy is not necessarily a sign of inhumanity, greed or lack of intelligence, but that for many, it is the only feasible reaction in the face of the disturbing magnitude of the climate crisis. Hope, on the other hand, can be the result of social pressures to stay positive (Head, 2016), is entangled with other conflicting emotions such as despair and should not be romanticized (Nairn, 2019).

I am particularly concerned with how paying attention to emotions can contribute to critical ECJ by discerning and critically evaluating pervasive yet often hidden discourses around the social justice aspects of climate action. The institutionalized disavowal of profane knowledge, as well as instrumentalized discourses around experiencing and managing emotions 'appropriately', places young people in a difficult agential deadlock. I suggest that paying attention to how emotions come to shape such discourses and circulate between the actors within them is crucial for addressing this deadlock in educational contexts.

A key example is the discourse around 'climate anxiety' (or 'eco-anxiety'). Generally defined as distress relating to the climate crisis, it can involve feelings of confusion, abandonment, moral injury, betrayal or pessimism (Marks et al, 2021). A recent study found that in children and young people globally, climate anxiety is widespread and linked particularly to a lack of adult and government responsiveness (Marks et al, 2021). Despite its widespread manifestation and dominance in public discourse around emotions and the climate crisis, I argue that climate anxiety is a problematic concept, and its uses and abuses depend on how reflectively and critically it is considered.

First, definitions of climate anxiety and its consequences and instrumentalizations are not neutral. While climate anxiety is generally found to be linked to feelings of uncertainty, unpredictability and uncontrollability, there is no consensus in the literature on whether it refers to existential anxiety, to strong or even pathological anxiety symptoms, or to feelings of worry and stress experienced as less severe (Pihkala, 2020). Several writers have warned against the tendency to pathologize climate anxiety, since it constitutes a 'reasonable and functional response to climate-related losses' (Cunsolo et al, 2020, p 261). Pathologizing climate anxiety can direct attention towards individual mental health rather than societal action (Clayton, 2020), involve rather crude and binary classifications of 'adaptive' or 'maladaptive' coping mechanisms (such as those that translate into action or paralysis), or aim to foster resilience rather than facilitate resistance (Taylor, 2020). Furthermore, when young people's justified emotions are reduced to anxiety and worrying, 'their capacities for sustained and critical thinking, and for responsible political action can be undermined' (Mayes and Hartup, 2021, p 19).

Second, climate anxiety (like other emotions), is unevenly distributed. In their research with 16–25 year olds in ten different countries, Marks et al (2021) found that young people in poorer countries and countries of the Global South expressed more worry, with a greater impact on their daily functioning. This corresponds to the fact that young people in these countries tend to experience greater climate change impacts, such as through extreme heat and droughts, floods, hurricanes or wildfires. In addition, there is a clear intergenerational dimension to the distribution of climate anxiety, with children and young people carrying the heavier burden. In our workshop series, this was expressed by many of the young people, for example:

> I hate it when people say 'Oh young people are so inspiring. How empowering it is to see you take action'. So why aren't you taking any action then? All these emotions, anger, frustration, worry – they come from: *we* need to do this. *You're* not going to do it. (Young climate activist, Scotland)

These emotions – anxiety, anger and frustration – point towards broader power dynamics in intergenerational relationships, particularly regarding children and young people's activism. For example, adults may not take young people's activism seriously or may distance themselves through patronizing attitudes and avoidance of personal involvement. These uneven distributions of climate anxiety (and other emotions) point towards a critical need for solidarity and empathy across generations and geographies as a fundamentally necessary emotional dimension of ECJ.

Third, climate anxiety has been critically interrogated for how it may mask underlying social justice dimensions. For example, Ray suggests that it is overwhelmingly white people who respond to the concept of climate anxiety, and who tend to dominate discussions around it: 'Is climate anxiety a form of white fragility or even racial anxiety? Put another way, is climate anxiety just code for white people wishing to hold onto their way of life or get "back to normal", to the comforts of their privilege?' (Ray, 2021, n.p.).

Positioning the climate crisis as humanity's biggest existential threat in history conveniently ignores the ongoing effects of colonialism and neocolonialism, and centres white people's often newly experienced ontological uncertainties. Additionally, Wray (2022) highlights how climate anxiety can be a driver behind ecofascist attitudes such as those espoused in the 'great replacement theory' (far-right conspiracies about white people being intentionally 'replaced' by people of colour). Ecofascist groups draw on environmentalist discourses without acknowledging historic power relations, and channel climate anxiety – for example, with regard to a loss of living standards or climate-related mass migration – into violent rhetoric and actions.

Retaining a critical perspective on climate anxiety as a discourse and what it may expose about broader social attitudes and injustices is thus crucial in terms of understanding the different and potentially harmful ways in which it can be used or misappropriated. At a time of 'disaster capitalism' (Klein, 2007), in which citizens are increasingly distracted by one crisis after the other while neoliberal and climate-damaging policies are implemented by governments, it is crucial to integrate a critical discourse around the emotions of these dynamics for ECJ. Approaching climate anxiety discourses critically, for example, can expose emotional investment in whiteness and its privileges, or how it is used by some people to undermine structural climate action and solidarity. By addressing the emotional dimensions of climate action in educational spaces, we can explore how emotions circulate, how they are claimed or attributed, weaponized or used to dismantle inequalities and privileges in relation to climate action, and thus ultimately fuel activism.

Profane knowledge and the problem of institutionalized disavowal

Callum: Marlies claims that emotion work is central to ECJ because it helps us to surface and critically evaluate 'pervasive yet often hidden discourses around the social justice aspects of climate action'. The problematization of climate anxiety that this entails involves the cultivation of emotional reflexivity. By 'emotional reflexivity', I mean a willingness to interrogate the roots and referents of such anxiety, to explore what it hides/makes visible and to position oneself in relation to such discourse. This is not easy or comfortable work, but it strikes me as a very productive way to think about how educators and learners, across a range of contexts, might use the ambivalent discourse of climate anxiety as a springboard into broader discussions about the intersection of climate action and social justice. Currently, it seems that the stronger valence for a discourse of anxiety is 'emergency' rather than 'justice'. The 'emergencification' of climate politics that we see globally, yet predominantly in the Global North (Ruiz-Campillo et al, 2021), can provide a supporting narrative for forms of ecofascism or green authoritarian populism justified by the need to 'do whatever it takes', with little regard for the ways in which climate action itself compounds existing social inequalities (Hulme, 2019; Wilson and Orlove, 2019). As Marlies alludes to in her concluding remarks, emergency can speak to a tacit desire to protect the privileges afforded to us by our class, race, legal status and gender.

Under such circumstances, I argue that education faces not so much a knowledge deficit problem, but rather a knowledge *disavowal* problem

(Haseley, 2019; Weintrobe, 2013). By 'disavowal', I mean 'the state of knowing and not knowing at the same time' (Haseley, 2019, p 110). One might, for example, disavow any serious antagonisms or contradictions between the desire to take climate action and the desire to protect one's white privilege (Ray, 2021; Wray, 2022). Moreover, one might disavow any inherent contradiction between climate action, growth economics and capital accumulation (O'Neill and Sinden, 2021). What can result from such disavowal is a fetishistic focus on reducing greenhouse gases as the horizon of our political lives (Hulme, 2019). Yet, climate anxiety need not be articulated in this way. Anxiety can be understood as the symptom of suppressed agency. In this way it can potentially be re-signified as part of the struggle for climate justice. My concern is the damage done when such processes of disavowal are institutionalized to the extent that young people are bombarded and burdened with apocalyptic narratives while being taught that we can avert disaster without systemic change.

In many respects, the policy context in Scotland is progressive. In addition to setting what are among the most stringent statutory targets in the world, the Scottish government has established an independent Citizens Assembly tasked with addressing how Scotland should tackle the climate emergency. This is in addition to an independent Just Transition Commission, whose vision is for a transition that seeks 'to capture opportunities to address existing inequalities, making urgent climate action a driver of positive change that improves wellbeing' (Just Transition Commission, 2021, p 38). Moreover, Scottish education policy has embedded a commitment to Learning for Sustainability (LfS), stating that all learners, throughout the life course, have an entitlement to LfS. It is important to locate this policy landscape in the broader context of youth-led climate activism in Scotland, with many young people framing their demands explicitly around climate justice.

In the ECJ workshops, a recurring theme that resonated with me was the disconnect that some young participants highlighted between an education system aspiring to centre LfS and the perception that it remains profane to critique growth economics in educational spaces. I would go so far as to argue that despite the ostensibly progressive policy context in Scotland, there is simply little visible appetite for questioning the Scottish government's mantra of 'sustainable economic growth'. The workshops highlighted an interesting dynamic whereby some young participants questioned the sustainability of economic growth, but identified a deficit in their own formal educational experience in this context. For example, I reflect here on a youth-led community conversation framed around the idea of education for a Just Transition. The conversation was composed of five young people of colour from youth work organizations in Edinburgh and Glasgow, and one young white activist from Teach the Future (TTF), a youth activist group campaigning for education reform to address the climate crisis. The six young people reflected on how they are not really educated about the sociopolitical or economic dimensions of climate change, and they also discussed how this lack of focus on systemic political, economic and social change in education only reinforced feelings of powerlessness and anxiety. This was because of the disjuncture between the evident scale of the problem and the kinds of individualized and reformist solutions promulgated in education and the media at large. As one young person campaigning for reform put it: "The education system is built to benefit those in power … it's about teaching us how to live in the world as it is, which is the capitalistic world based on colonialism" (young climate activist, Scotland).

The young people in this conversation discussed schooling as an instrument of socialization, addressing the ways in which, in their experience, schooling was ultimately about reproducing a capitalist workforce. They also eloquently discussed the need to decolonize the curriculum and discussed the lack of diversity in the youth-led climate movement. Rather than learning about these issues in school, young people learned together through their own action and educated themselves about the connections between unsustainability, capitalism and colonialism through social media. For example, another young participant reflected that:

> We were briefly taught about capitalism in Modern Studies when I was in secondary school but it was more like capitalism is the way forward and communist countries are so backwards … Now I've self-taught myself from social media and hearing other people's experiences and perspectives. It's crazy that I've learnt … from social media that capitalism is actually destroying people's lives, it's destroying the planet, maybe it's not the way forward. (Young climate activist, Scotland)

This conversation led into a panel discussion – chaired by a young activist – which addressed the question of how we can make space to collectively imagine postgrowth futures in climate education. An interesting point of contrast here for me was the difference between the education system in Scotland and in England. Researcher and activist Dena Arya highlighted the revanchist movement in English education policy, citing the Department for Education's statutory guidance for schools and teachers in 2020, which restricts students' access to 'anti-capitalist' material:

> Schools should not under any circumstances use resources produced by organisations that take extreme political stances on matters … Examples of extreme political stances include, but are not limited to: a publicly stated desire to abolish or overthrow democracy, capitalism, or to end free and fair elections. (Department for Education, 2020, n.p.)

This then represents a much more overt institutionalized disavowal of 'profane' knowledge, where young activists' efforts to reform education policy are ensnared in the ideological 'war on woke'. This is not lost on young climate activists in TTF, who offered a scathing assessment of this approach in their response to the Department for Education's draft Climate and Sustainability Strategy:

> Whilst the strategy includes recognition of student leadership, it also states: 'it would not be appropriate to encourage pupils to join specific campaigning groups or engage in specific political activity, such as protests' ... it feels dismissive of the youth climate movement ... *Campaigns also have an important role in developing critical thinking skills, agency and they tend to be interdisciplinary, all of which should be acknowledged rather than dismissed.* (Teach the Future, 2021, p 3, emphasis added)

This final sentence from TTF is central to our collective position on ECJ because justice itself is partly about epistemic agency. Indeed, for the young people involved in our workshops, their critical deliberations about the contradictions of sustainable growth were informed by their own activism and self-directed learning on social media, as illustrated earlier.

Granted, the Scottish policy context is different from the English context. However, a question deserving further empirical research is whether there is any serious attempt, in practice, to confront both the fetish of growth and the disavowal required to maintain commitments to sustainability and capitalist growth economics. For this, we need to look to the demands, analyses and, ultimately, learning and public pedagogies generated by the vibrant extraparliamentary politics of current social movements. Yet educators in schools and communities also need to be confident enough to engage in these open-ended discussions by developing fit-for-purpose curricula. The basis for such curricula already exists; ecological economics is awash with ideas around steady-state economies, zero-growth economies and degrowth (see, for example, Hickel, 2021; Jackson, 2021; Schmelzer et al, 2022), which might be considered alongside contemporary arguments for healthy green growth (for example, Stokes, 2022). Indeed, proponents of degrowth such as Schmelzer et al (2022, p 207–209) recognize that to redesign institutions and infrastructures reliant on growth, we must redesign 'mental infrastructures' through educational institutions.

What strikes me as important here is the need to recognize that although we live under capitalism, it doesn't follow that we have a grasp of what is really going on 'under the hood', beyond the ideological clichés that we absorb. Around 50 years ago, the infamous Club of Rome report 'Limits to Growth' modelled that perpetual economic growth based on increasing use and depletion of finite natural resources would lead to ecological and economic disaster. Arguably, the ecological critiques of growth are de facto anti-capitalist because one of the intrinsic features of capitalism is its systemic requirement for self-expanding value and compound growth (Fraser, 2014; Harvey, 2017). A zero-growth capitalist economy is a contradiction in terms because '[f]or all capitalists to realise a positive profit requires the existence of more value at the end of the day than there was at the beginning. That means an expansion of the total output of social labour. Without that expansion there can be no capital' (Harvey, 2017, p 320). To understand how and why capitalism relies on 'nature' as a 'free gift', while disavowing its own destabilization of climate and ecosystems, it is necessary for us to look at it ecologically (Fraser, 2021). Therefore, I hold that in order for the Scottish policy architecture of LfS to live up to its espoused values, it has a responsibility to make space to explore the urgent and legitimate questions that young people are posing. Certainly, there is a strategic discussion to be had over how such conversations and curricula ought to be framed, but to suggest that we ought to protect young people on the grounds of either anxiety or indoctrination is disingenuous, patronizing and negligent. As Verlie and Flynn (2021, pp 4–5) astutely highlight, if educational institutions have a duty to protect children and young people, then they fail in this duty through acts of 'protective silencing', which stifle imagination and critical thought.

Education: policy, politics and pedagogies

Beth: Callum briefly introduced the Scottish LfS policy context. The details of this policy architecture make clear the opportunity that exists to develop ECJ as part of a broader curriculum of sustainability. Yet, while this seemingly accommodating policy context exists –and is currently undergoing a period of revision – it has been harder to translate the ambitious vision into practice. In my view, the practice of LfS across primary and secondary schools has been patchy and inconsistent. I work closely with education professionals, including teachers (primary and secondary), policy makers and those working within educational governance. I work in an initial teacher education institution within a School of Education and participated in the ECJ workshops described earlier, learning alongside our collaborators. I am also a planetary citizen experiencing the consequences and crises of the twin planetary emergencies of biodiversity collapse and climate catastrophe in ways nuanced to the lifeworld I inhabit. From this partial perspective I recognize the issues raised by Marlies, and I am witness to the contradictions described by Callum. As I reflect upon my co-authors' points, I return to the conversations I have had with teachers. They have shared the barriers and challenges they face when trying to create educational opportunities aligned with LfS Scottish policy agendas that seek, perhaps implicitly, to overcome such disavowal. I hear the internalized tensions, felt both personally and professionally, between a desire to explore and discuss what they want, and the constraint of what they feel able to do within the culture and context of their school environment. From this stance, I offer three observations.

First, there is incremental progress in relation to the development and embedding of LfS policy across and between educational spaces in Scotland, ranging from governance to higher education, secondary and primary school contexts. Patchy provision is related to unclear terminology and implementation, which in turn is related to a lack of governmental investment, support and resourcing to create the time and space needed to fully consider how policy may be contextualized, implemented and delivered. Currently there are interrelated educational priorities taking precedence, such as the focus on a post-COVID-19 recovery curriculum to address the increasing attainment gap, as well as national education discussions which are considering national assessment and inspection processes (Education Scotland, 2022). LfS, while intrinsically related to these discussions, is one of a series of competing policy priorities. However, the tension between the slow progress of policy and the urgency of the sustainability crises we face are felt beyond education in the wider civil society. For example, the youth-led movement Teach the Future (TTF) (2020, n.p.) argues that despite 'admirable policy commitments ... unfortunately our education system routinely fails to educate, prepare and equip us, and our fellow students, to abate and stop climate emergency and ecological crisis to deliver climate justice'. This claim only serves to highlight that policy entitlement to LfS is not widely known by students and teachers, and is often not regarded as a priority. There is no stated or established pathway for LfS through Scottish education, and exam performance in siloed subjects is still held up as the primary indicator of 'success'.

Second, policy and formal curriculum inclusion alone is not enough. Curriculum making requires a collaboration between teacher and learner (whoever is occupying which role at a given time) and the inclusion of LfS policy rhetoric within this negotiation will be guided by the individual's knowledge and understanding of that policy intention, but also by the context and curiosity of those in that educational space. This point is not revelatory; it is made clear by young activists involved with TTF, as is evident in the current research findings, and often arose in conversations during our workshops. However, I raise it again as I was acutely aware that there were teachers and youth workers participating in the workshops who were actively working in schools and communities with young people to create spaces to discuss and act on climate injustice. In those cases, they were using LfS policy to prise open debates about the nature and purpose of education, drawing on the moral and ethical purpose of teaching and learning, and examining the critique of schooling as forming capitalist mental infrastructures through competitive performance-orientated systems (Schmelzer et al, 2022). During the workshops, broad statements were made about what 'education' should do and what teachers did not do, but ought to. How did those teachers in the audience feel when they heard education being identified as both the dishonourable protagonist and the holy grail?

Striving for inclusion within policy and formal curriculum does not ensure translation into professional practice (Evans et al, 2021). The principle of co-creation – which interestingly is at the heart of LfS policy and much of sustainability education – requires a curriculum to be developed as a collaborative process between 'student' and 'teacher' based on a degree of freedom to 'negotiate content and method' (Bron et al, 2022, p 39). Additionally, sustainability education requires space to imagine anew, which if taken seriously asks us all (teacher and learner) to think beyond the particular educational, political and societal systems in which we exist and to bring a newness to the world (Postma and Smeyers, 2012, p 403). Tensions arise when we encourage futurist thinking, yet tether those possibilities to an education system bound by mechanistic notions of performativity and a society driven by the imperatives of capitalist growth (Escobar, 2018). In this context, as we have argued earlier, authentic co-creation involves recognizing youth-led movements as educative spaces in their own right (McGregor and Christie, 2021; Verlie and Flynn, 2022). Equally, however, it entails reckoning with the anxiety that educators themselves experience as they try to exercise relative agency in the context of broader structural constraints.

The third point arises from a deeper conversation with one educator who reflected on the contradictory space in which they found themself as someone who was invested in developing LfS within their teaching practice, but found challenges in doing so. They raised two tensions. First, they referenced a professional tension between teacher agency and curricular structure: identifying "not enough hours in a day" and a familiarity in being "used to teaching the curriculum" that seemed to prohibit their freedom to explore issues and develop the curriculum in their own way. This is interesting given the nature of Curriculum for Excellence (Scotland's curriculum) and LfS policy as enabling and supportive of personalization and co-creation between teacher and learner. This reflects a growing concern that particular systems of education, more broadly, are in a state of crisis (Santos, 2017). If we are to take climate anxiety seriously in education for sustainability, then we must be cognisant that the educational structures, processes and systems may contradict with the very nature and purpose of the education we strive for. Verlie and Flynn (2022, p 3), examining the school strikes for climate in Australia, capture this point in their shrewd observation that 'students have felt that walking away from school – literally and symbolically – is the most agentic thing they could do for ecological justice'. This reckoning has arrived within Scottish education where young activists demand authentic and reflexive education that calls into the question its own presuppositions (TTF, 2020). The second tension relates to one of our workshops on the theme of 'Education for a Just Transition'. By 'Just Transition', we mean a socially just transition away from an extractivist economy that does not abandon communities whose livelihoods depend on it. The teacher worked in an area where many pupils' parents were employed directly or indirectly by the oil industry and they recognized that what we 'know' can be disavowed when it rubs up against the realities of work and everyday life. For example, a young Indigenous activist from Louisiana in the US gave a keynote speech during our ECJ workshops in which she reflected that:

> People where I live see us talking about climate change as us directly going against the families working in fossil fuel industries, so it's very hard to talk about. [The community] knows the reason why coastal erosion is happening but it is so taboo to talk about. I can't even imagine talking about it in school.

When the teacher thought about similar issues in terms of their professional practice in Scotland, they could see how such discussions in the classroom could bring pupils into personal conflict where issues of right and wrong go beyond scientific facts into political, ethical and emotional dimensions. They felt uncomfortable, unqualified perhaps, to stray from prescribed curriculum into areas that could provoke emotional responses. If we enter into sterile versions of sustainability education for fear of troubling and uncomfortable conversation, then we risk perpetuating hegemonic discourses within education rather than disrupting them. Teachers and youth workers therefore need the space to explore their own emotions, anxieties and ideological assumptions alongside children and young people. The discourse of climate anxiety could be useful in this context, provided that it is understood partially as the legitimate expression of structural constraints, both within educational institutions and the wider community.

Conclusion

In this chapter, we have critically reflected on our involvement in a series of ECJ workshops developing a critical analysis of climate anxiety as a hegemonic emotional discourse, circulating in activist and educational spaces. Rather than merely deconstructing the discourse of climate anxiety, we have argued throughout that it might be re-signified in ways that articulate more directly with issues of social injustice. Moreover, we have suggested that a starting point for such critical educational engagement is to recognize anxiety as a symptom of the disavowal of profane knowledge: young people and educators are asked to confront what is, for all intents and purposes, an apocalyptic discourse within the parameters of the status quo. Young people and education are often constructed as panaceas in public discourse in ways that allow others to distance themselves from addressing their own complicity. Yet public discourse around young people and education is contradictory. Young people are simultaneously 'the answer' and dismissed as idealistic or in need of paternalistic protection from indoctrination. Moreover, education is not just 'the solution' but also 'the problem' when it is represented by the populist Right as the breeding ground of extreme leftists and the 'wokerati', as witnessed in public concerns over the teaching of a whole range of 'contentious' issues, including the climate crisis, Trans Rights and structural racism.

This contradictory discourse manifests in overt and much more subtle ways. In our case, we have discussed how the English policy context provides an overt example of what Verlie and Flynn (2021) call protective silencing, while the Scottish policy context is ostensibly more progressive. However, being more progressive in policy can make the problem more insidious and difficult to address. Can we learn for sustainability in a context that has no appetite for questioning the arch policy priority of sustainable economic growth? Protective silencing may have deleterious consequences for the mental health of young people as well as educators tasked with enforcing it. The argument that we have made is that this is a driver of climate anxiety because it stifles agency. More specifically, it stifles the exercise of what Emirbayer and Mische (1998) call 'projective agency', meaning the agency to imagine possible futures together and think beyond received wisdom. We would like to conclude by offering four practical suggestions for education in schools and communities that follow from our analysis.

First, climate anxiety can be the starting point for reflexive discussions of intragenerational and intergenerational injustice in several ways. Educators and learners might use it as a way into intergenerational dialogues about the disproportionate burden that young people face and the contradictory discourses that position young people as saviours of the future and in need of protection from explorations of alternative futures. Moreover, educators and learners might use it as a way into exploring and unpacking the roots and referents of anxiety beyond climate. This involves difficult discussions of who stands to gain and who stands to lose out in climate action. It involves conversations about social injustice and the relative loss of privilege.

Second, education for climate justice must recognize agency itself as a central axis of injustice where this recognition has at least two practical implications. Young people's climate activism must be understood as educative in its own right and as a legitimate starting point of, and driver for, sustainability education (McGregor and Christie, 2021; Verlie and Flynn, 2021). It is through collective action that young people learn about the social organization of power, how to build solidarity and work with others, articulate demands and educate themselves about the relationship between climate crisis, capitalism and coloniality – the domain of what we have called 'profane knowledge'. Yet we cannot simply assume justice within activist and educational spaces: to explore justice is an educational task that involves asking whose voices are represented within activist communities and educational institutions, and with what consequences for how climate action is conceived.

Third, and in relation to the preceding point, education should extend beyond the classroom not only into activist spaces but also into the everyday life of households and communities 'locked into' particular patterns of work and consumption (Shove, 2010). Thus, education for climate justice must always be education for a Just Transition. Such education would recognize that although the transmission of facts and moralizing pleas for behaviour change may be politically palatable, they are not going to be effective in engaging and mobilizing those communities at the sharp end of green policies that fail to address their material living conditions. There is arguably an important role for community-based youth work and adult education in this context, which has the potential to unlock avenues for action by reframing toxic or stalemate political debates through solidarity and planetary justice.

Finally, to the extent that education for climate justice is a process of co-construction, teachers and youth workers must be given space to explore their own anxieties alongside young people. This would involve finding ways to address rather than gloss over the structural constraints shaping their practice within schools and communities and the contradictory policy discourses that they are tasked with navigating.

Notes

[1] More information, including resources and the final report, can be accessed here: https://www.scottishinsight.ac.uk/Programmes/UNGlobalGoals/EducationforClimateJustice.aspx.

[2] While a full exploration of our own positionalities is beyond the scope of this chapter, writing as white people of different European nationalities working in the UK higher education context compels us to interrogate relations of power from this perspective.

References

Ahmed, S. (2014) *The Cultural Politics of Emotion*, 2nd edn, Abingdon: Routledge.

Bron, J., Bovill, C. and Veugelers, W. (2022) 'Students experiencing and developing democratic citizenship through curriculum negotiation: the relevance of Gath Boomer's approach', *Curriculum Perspectives*, 42(1): 39–49.

Clayton, S. (2020) 'Climate anxiety: psychological responses to climate change', *Journal of Anxiety Disorders*, 74: 102263.

Cunsolo, A., Harper, S.L., Minor, K., Hayes, K., Williams, K.G. and Howard, C. (2020) 'Ecological grief and anxiety: the start of a healthy response to climate change?' *The Lancet Planetary Health*, 4(7): e261–e263.

Department for Education (2020) 'Political impartiality in schools', UK government [applies to England], Available from: https://www.gov.uk/government/publications/political-impartiality-in-schools/political-impartiality-in-schools [Accessed 3 November 2022].

Education Scotland (2022) 'Covid-19 education recovery', Scottish government, Available from: https://www.gov.scot/groups/covid-19-education-recovery-group/ [Accessed 21 October 2022].

Emirbayer, M. and Mische, A. (1998) 'What is agency?', *American Journal of Sociology*, 103(4): 962–1023.

Escobar, A. (2018) *Designs for the Pluriverse: Radical Interdependence, Autonomy and the Making of Worlds*, Durham, NC: Duke University Press.

Evans, N.S., Inwood, H., Christie, B. and Ärlemalm-Hagsér, E. (2021) 'Comparing education for sustainable development in initial teacher education across four countries', *International Journal of Sustainability in Higher Education*, 22(6): 1351–1372.

Fraser, N. (2014) 'Behind Marx's hidden abode', *New Left Review*, 86: 55–72.

Fraser, N. (2021) 'Climates of capital', *New Left Review*, 127: 94–127.

Gagen, E.A. (2015) 'Governing emotions: citizenship, neuroscience and the education of youth', *Transactions of the Institute of British Geographers*, 40(1): 140–152.

Harvey, D. (2017) *Seventeen Contradictions and the End of Capitalism*, London: Profile.

Haseley, D. (2019) 'Climate change: clinical considerations', *International Journal of Applied Psychoanalytic Studies*, 16(2): 109–115.

Head, L. (2016) *Hope and Grief in the Anthropocene: Re-conceptualising Human–Nature Relations*, Abingdon: Routledge.

Hickel, J. (2021) *Less Is More: How Degrowth Will Save the World*, London: William Heinemann.

Hulme, M. (2019) 'Climate emergency politics is dangerous', *Issues in Science and Technology*, 36(1): 23–25.

Jackson, T. (2021) *Post-growth Life after Capitalism*, London: Polity Press.

Jones, C.A. and Davison, A. (2021) 'Disempowering emotions: the role of educational experiences in social responses to climate change', *Geoforum*, 118: 190–200.

Just Transition Commission (2021) *Just Transition Commission: A National Mission for a Fairer Greener Scotland*, Scottish government, Available from: https://www.gov.scot/publications/transition-commission-national-mission-fairer-greener-scotland/pages/5/ [Accessed 28 March 2023]

Klein, N. (2007) *The Shock Doctrine: The Rise of Disaster Capitalism*, London: Macmillan.

Marks, E., Hickman, C., Pihkala, P., Clayton, S., Lewandowski, E.R., Mayall, E.E., Wray, B., Mellor, C. and van Susteren, L. (2021) 'Young people's voices on climate anxiety, government betrayal and moral injury: a global phenomenon', *The Lancet*, Available from: https://ssrn.com/abstract=3918955 or http://dx.doi.org/10.2139/ssrn.3918955.

McGregor, C. and Christie, B. (2021) 'Towards climate justice education: views from activists and educators in Scotland', *Environmental Education Research*, 27(5): 652–668.

Mayes, E. and Hartup, M.E. (2021) 'News coverage of the School Strike for Climate movement in Australia: the politics of representing young strikers' emotions', *Journal of Youth Studies*, 25(7): 994–1016.

Nairn, K. (2019) 'Learning from young people engaged in climate activism: the potential of collectivizing despair and hope', *Young*, 27(5): 435–450.

Norgaard, K.M. (2011) *Living in Denial: Climate Change, Emotions, and Everyday Life*, Cambridge, MA: MIT Press.

O'Neill, K. and Sinden, C. (2021) 'Universities, sustainability and neoliberalism: contradictions of the climate emergency declarations', *Politics and Governance*, 9(2): 29–40.

Pihkala, P. (2020) 'Anxiety and the ecological crisis: an analysis of eco-anxiety and climate anxiety', *Sustainability*, 12(19): 7836.

Postma, D. and Smyers, P. (2012) 'Like a swallow, moving forward in circles: on the future dimension of environmental care and education', *Journal of Moral Education*, 41(3): 399–412.

Ray, S.J. (2021) 'Climate anxiety is an overwhelmingly white phenomenon', *Scientific American*, Available from: https://www.scientificamerican.com/article/the-unbearable-whiteness-of-climate-anxiety/ [Accessed 20 June 2022].

Santos, B. (2017) *Decolonising the University: The Challenge of Deep Cognitive Justice*, Newcastle: Cambridge Scholars Publishing.

Schmelzer, M., Vetter, A. and Vansintjan, A. (2022) *The Future is Degrowth: A Guide to a World beyond Capitalism*, London: Verso.

Shove, E. (2010) 'Beyond the ABC: climate change policy and theories of social change', *Environment and Planning A: Economy and Space*, 42(6): 1273–1285.

Stokes, P.E. (2022) *Tomorrow's Economy: A Guide to Creating Healthy Green Growth*, Cambridge, MA: MIT Press.

Taylor, S. (2020) 'Anxiety disorders, climate change, and the challenges ahead: Introduction to the special issue', *Journal of Anxiety Disorders*, 76: 102313.

Teach the Future (2020) *Scotland Policy Asks*, Available from: https://uploads-ssl.webflow.com/5f8805cef8a604de754618bb/5fa3e6b5d6845067a6eeadfd_Asks%20Scotland.pdf [Accessed 15 September 2022].

Teach the Future (2021) *Teach the Future's Response to the Department for Education's Draft Climate and Sustainability Strategy*, Available from: https://www.teachthefuture.uk/blog/response-to-the-dfes-draft-climate-and-sustainability-strategy [Accessed 3 November 2022].

Verlie, B. and Flynn, A. (2022) 'School Strike for Climate: a reckoning for education', *Australian Journal of Environmental Education*, 38(1): 1–12.

Weintrobe, S. (2013) *Engaging with Climate Change*, Abingdon: Routledge.

Wilson, A.J. and Orlove, B. (2019) 'What do we mean when we say climate change is urgent?', *Centre for Research on Environmental Decisions Working Paper 1*. New York: Centre for Research on Environmental Decisions, Available from Academic Commons, Columbia University Libraries: https://academiccommons.columbia.edu/doi/10.7916/d8-b7cd-4136 [Accessed 3 November 2022].

Wray, B. (2022) 'Eco-fascism is part of climate anxiety', *Gen Dread Newsletter* [online], 18 May, Available from: https://gendread.substack.com/p/eco-fascism-is-part-of-climate-anxiety [Accessed 20 June 2022].

12

White Audacity and Student Climate Justice Activism

Natasha Abhayawickrama, Eve Mayes and Dani Villafaña

Introduction

In recent years, school students have gathered in mass mobilizations across the world to call on political leaders to take urgent action on climate change and to respond to its inequitably distributed effects. Since 2018, large numbers of Australian school students have walked out of school to protest as part of global youth climate actions, known as School Strike 4 Climate (SS4C) in Australia. For many years before these highly mediatized mass strike actions, in communities where environmental destruction intersects with colonialism, capitalist exploitation and racism, young people have long played vital roles in caring for Land/Country (on these terms, see Chapter 2). Young people have been an integral part of intergenerational political struggles for Indigenous sovereignty and land rights, although this activism has not necessarily been recognized as 'activism' by mainstream environmental movements (Ford and Norgaard, 2020).

In the context in which we write, on these lands and waters currently known as Australia, grassroots youth-led climate networks have explicitly foregrounded 'climate justice' in their political demands. In the Australian context, 'climate justice' foregrounds resourcing 'Aboriginal and Torres Strait Islander-led solutions that guarantee land rights and care for Country' (SS4C, 2021). However, climate justice is an evolving and contested concept, meaning, as Jafry et al write, 'different things to different people' and 'different things to the same people depending on a particular time and space' (2018, p 3). The concept of climate justice emerges from First Nations land and water rights, and environmental justice and ecofeminist movements. Those calling for climate justice foreground First Nations sovereignty and the inequalities experienced between industrialized countries in the Global North and the Global South, gender injustices, environmental racisms and multispecies injustices (Whyte, 2020; Sultana, 2021). Yet, in Australia (as elsewhere), the climate justice movement carries with it the histories of environmentalisms' dividing lines of exclusion – white privilege, class privilege and able-bodied privilege. In addition, there are tensions and uneasy alliances between 'Green' and 'Black' climate activisms (Vincent and Neale, 2016). As we consider in this chapter, there may be new strategic alliances, friendships and intimacies formed in striking together; there may also be uncomfortable politics of solidarity (Land, 2015) when young people come together, across embodied differences, to organize strikes for the climate.

There has been burgeoning sociological interest in youth climate justice activism across the world, particularly in the Global North. This work has drawn attention to the collective hope (Nairn, 2019), joy and 'radical kindness' that form and move in and through youth climate justice activist networks and communities (Pickard, 2022; Bowman et al, 2023). We acknowledge and have experienced first-hand the affirmative emotions associated with youth climate justice organizing spaces. Ruchira Talukdar, a researcher-activist working across Australia and India and co-founder of Sapna South Asian Climate Solidarity (also a contributor to this book and this project team's mentor; see Chapter 6), notes that youth-led climate justice networks are 'more likely to be racially diverse than mainstream climate movements' (2022, p 7). Talukdar describes how youth-led climate justice movements frame climate change not just in scientific terms – as is the tendency of mainstream environmental and climate organizations in the Global North – but also draw attention to 'the colonial and racist underpinnings of the current climate crisis' (2022, p 7). We agree that youth-led climate justice networks are more diverse and attuned to intersectional experiences of injustices than mainstream climate groups.

However, while acknowledging the strengths of youth climate justice activism, this chapter seeks to further unpack and interrogate the intersectional dynamics and rage at injustices that circulate within and through youth climate justice organizing spaces. These micropolitical dynamics and politics of emotion are tangled up with feelings of hope, joy and kindness, and are differentially experienced within social movements (see for example, Ahmed, 2004). In this chapter, we elaborate on a concept forged through experience – a phrase that Natasha initially articulated in a research team meeting as 'white audacity'. 'White audacity' names a mode of presumption to power contoured by whiteness that has negatively impacted on youth-led climate organizing. We also draw attention to compelling moments where white audacity has been disrupted, exposed and challenged. We conclude by looking laterally at the collective work of Seed Mob and Pacific Climate Warriors; we also call for further reflection on the organizing model of youth-led climate justice networks in order to collectively nurture strong intersectional climate justice work.

The authors of this chapter, Natasha Abhayawickrama (she/her) and Dani Villafaña (she/her), are graduated school-student organizers for School Strike 4 Climate (SS4C), writing with Eve Mayes (she/her), a university-based educational researcher. Natasha and Dani were organizers with SS4C in 2019–2022 (Natasha) and 2018–2021 (Dani) while living on Dharug Country in Western Sydney. Both Natasha and Dani now both live on Gadigal Country, in the inner west of Sydney. Eve is a second-generation white settler scholar who worked as a secondary school English teacher on Dharug Country, in Western Sydney, for ten years (2007–2016), and lived on Gadigal Country before she moved to Wadawurrung Country, Victoria, where she currently lives. This chapter narrates and analytically reflects on the dynamics within the student-led organizing circles of School Strike 4 Climate, with particular attention to the operations of whiteness (see later on in this chapter) in youth climate justice organizing spaces within the metropolitan city of Sydney, Australia, where the three co-authors live (Natasha and Dani) and have previously lived (Eve). We contextualize our place-based positionalities here to make clear that we are writing about our lived experiences of the within-movement politics of solidarity situated in one city (Sydney), focusing on 2019 and 2020.

Rationale

It is important to explicitly articulate the political purpose of writing about these within-movement and within-city politics of solidarity. While the climate crisis is a 'deep-seated systemic crisis', necessitating transformative structural change, movements 'striving to create such change are inevitably shaped and limited by the same structural conditions they seek to dismantle' (Simpson and Choy, 2023, p 1). Activist spaces are 'not hermetically sealed spaces' (Choudry, 2015, pp 11–12); oppressive relations and practices can also show up in ostensibly 'progressive' movements and activist spaces (Gorski and Erakat, 2019). Sexism, classism, racism, homophobia, transphobia, able-ism and fascist tendencies, when understood intersectionally, are 'mutually constitutive, not conceptually distinct' (Hancock, 2016, p 71). Social movement scholarship has been critiqued for a tendency towards abstraction and universalism, written from the 'outside' with an empirical focus on activist work in the public sphere (Choudry, 2015, p 57). This chapter responds to calls for sustained analysis of internal movement dynamics, particularly in relation to whiteness and racialization. We take this focus because the impact of these negative experiences may 'debilitate activists' abilities to remain engaged' and thus pose a 'formidable threat to the sustainability' of movements (Gorski and Erakat, 2019, p 784).

The relative silence of those positioned in the academy on internal social movement dynamics contrasts with the concerns of movement organizers. Within activist circles in Australia, conversations about the dynamics of gender, white supremacy, racism and classism, as they shape and constrain solidarity, have long been on organizers' agendas, especially in conversations about 'white allies' and First Nations struggles (Foley, 2000; Land, 2015; see also Chapter 9). Beyond Australia, there have been important critiques of climate and global justice campaigns by Black, Indigenous and People of Colour (BIPOC) activists – for example, see the archived resources compiled by the Colours of Resistance (n.d.) in the US and Canada between 2000–2006 to develop anti-racist politics in the movement against global capitalism, and the grassroots collective Wretched of the Earth's open letter to Extinction Rebellion (2019).[1] Sapna's report (Talukdar, 2022) brings together the experiences, shared in interviews, of 12 South Asian young people based in Melbourne and Sydney who had been involved in climate activism between 2016 and 2019. Talukdar argues that sharing the experiences of young South Asian climate activists 'could generate vital cross-cultural learning to decolonise and diversify climate activism', enabling campaigns to 'connect the climate crisis to multi-generational South Asian communities through telling the climate story in culturally relevant ways' (2022, p 10). This chapter seeks to complement the stories and insights shared in the Sapna report, noting that one of the authors of this chapter was an interviewee for this report. We urge readers of this chapter to also read the Sapna report, which details more specific examples of the dynamics that we will outline subsequently. We are unromantic about youth climate justice activism, but still believe that solidarities are possible.

In this chapter, we inquire into the conditions under which problematic movement dynamics are (re)produced and consider how these conditions may be reshaped. We seek to learn from, and to encourage further reflection on, moments of failed solidarity, and to question what a more intersectional organizing model might look like. This chapter seeks 'to be of use' (McKenzie, 2009; Fine and Barreras, 2001) to current critical conversations and actions about whiteness, intersectionality and racial justice taking place in heterogeneous climate justice activist circles both locally and globally. Intervening into problematic internal dynamics in youth-led climate justice organizing spaces is one way to grow strong intersectional climate justice organizing networks that prefigure alternative ways of living and relating together and with the Earth (Escobar, 2022).

Co-authoring as conversation

This chapter is a dialogue between three members of a larger research team[2] on a project exploring young people's climate justice activism: four of the research team members are 19–22 years old and active climate justice organizers, and are involved in this research as a part-time job. This research team is working together on an Australian Research Council project (project number DE220100103) co-constructing accounts of students' climate justice activism(s). Differentially positioned across identity markers and embodied experiences, we dialogue across our respective knowledges and embodied experiences in and across organizing and academic spaces. Our approach is informed by feminist critical participatory activist traditions that recognize that 'expertise is widely distributed but legitimacy is not; that those who have experienced injustice have a particularly acute understanding of the affects, capillaries, consequences, and circuits of dispossession and privilege' (Fine, 2018, p 80). What follows is deliberately kept in dialogical form, as one strategy to resist the tendency for academic work on social movements to come across as the primary site of research and knowledge production. We agree with Choudry on the need for 'knowledge about movements to be produced in a genuine dialogue with activists and the knowledge that they produce themselves', in an 'interchange' that 'attends to concerns about race, gender, and class dynamics within organizing contexts' (2015, p 61).

Prior to writing this chapter, the research team had already been meeting fortnightly over a six-month period, preparing the project's design, engaging in consultation processes with members of the project's First Nations Critical Reference Group and Stakeholder Reference Group, and preparing an institutional ethics application. The intention was to shape the project design to be accountable to those directly involved in climate justice organizing and activism. The team co-developed and practised 'research conversations' (the central project method) during these meetings: sharing stories reciprocally and adding to others' stories. Bevington and Dixon, writing about 'movement-relevant theory', pose the following questions: 'What issues concern movement participants? What ideas and theories are activists producing?' (2005, p 198). Our process of writing has sought to elaborate on the issues, ideas and theories that research team members have *already* been generating in their own activist praxes before this research team came together. In preparing this chapter, we wrote notes into a shared Google document, with Natasha taking the lead in writing, and later reshaped these stories into a dialogical structure. Where multiple names are written at the beginning of a paragraph, this indicates collaborative work on the paragraph in the Google doc.

At the same time as we describe this collective writing process, and our intention to craft more reciprocal dialogical accounts of experiences, it is important to foreground the tensions and ambivalences that are marbled through this co-authoring work. While all authors are paid members of a research team, Eve Mayes is the 'principal investigator' on the project and manages the project budget (and its finite number of employed personnel hours). Our extensive involvement in activist circles, embodied knowledges, age, intersectional identities, experiences of racialization and prior academic writing (at school and university and in academic circles) differentially shape our respective sense of confidence in writing together. There are always micropolitical forces in writing with others: who can add, modify and delete texts? Who has time and headspace to write? These pressures are amplified within the intergenerational institutional dynamics of an academic research team.

The Master's House and Youth Climate Justice Organizing

Natasha: In this chapter, we work with the movement training framework of 'The Master's House'. In 2021, in my local Sydney School Strike 4 Climate group, following some of the challenges that we describe below, we began running 'introduction to climate justice' workshops for organizers, facilitated by student organizers of colour. Three other organizers and I created a PowerPoint, drawing on what we had learnt in other forums, and from resources shared by other organizers before us; for example, in the online 'Strike School' workshops in 2020 facilitated by Original Power, and from climate justice resources from the Australian Youth Climate Coalition and Democracy in Colour. Within this training, students learned about 'the Master's House'[3] (see Lorde [1984], which will be discussed later on). This is a visual metaphor of a house (the structure of domination) held up by three foundational pillars: capitalism, white supremacy, and patriarchy. This training included discussion of how corporations and specifically, fossil fuel companies and corporations, are shaped by and benefit from these pillars of oppression.

These pillars are also part of activist spaces themselves; activist group meetings are not pure progressive spaces that exist beyond structures of oppression. Just because a person is a school student does not mean that they are untainted by structures of power and domination, especially when these have come to shape the privileges and inequalities that they benefit from each day. Youth activist spaces are sites where colonial, capitalist, and patriarchal logics still circulate, structure and shape what is possible and not possible to be said and done.

Eve: While there is much to say about capitalism and patriarchy (see Villafaña et al, 2023), this chapter focuses on one pillar of oppression: white supremacy, and its subtle workings in even 'progressive' youth climate justice spaces. In Australia, Ghassan Hage (1998) has explained the historical construction and contingency of whiteness and its 'fantasy position' of dominance in

Australia, from European colonial invasion and expansion to its operation in policing the borders of inclusion in civic life. Goenpul scholar Aileen Moreton-Robinson writes that whiteness operates 'as race, as privilege and as social construction': as 'the invisible omnipresent norm' and the 'standard by which certain "differences" are measured, centred and normalised' (2020 [2000], pp xviii, xix).

Whiteness is more than essence and skin pigmentation. Whiteness, and access to its material and symbolic benefits, can be accumulated through cultivating a particular 'linguistic, physical and cultural disposition' characteristic of 'Anglo-ness' (Hage, 1998, p 53). As Michele Lobo writes, one can accumulate whiteness by 'adopting the look, accent, taste, attitudes, behaviour and lifestyle preferences that conform to the norms set by Anglo-Australians' (2010, p 102). Whiteness is, as Arathi Sriprakash and colleagues write, 'mediated through ableism, hetero-patriarchy and classism as simultaneous forces of domination under settler colonialism', and 'whiteness works differently upon different racialised groups: Indigenous; white settler or immigrant; Black or Brown immigrant, refugee or settler' (Sriprakash et al, 2022, p 5).

White supremacy is entwined with the climate crisis. In 2014, with the emergence of #BlackLivesMatter, Naomi Klein wrote that the 'whispered subtext of our entire response to the climate crisis' is 'an economic order built on white supremacy', and that this subtext 'badly needs to be dragged into the light' (2014, para 4). Ghassan Hage has described racism as an 'environmental threat' in a way that resonates with the Master's House framework used within youth climate justice networks: 'Racism is an environmental threat because it reinforces and reproduces the dominance of the basic social structures that are behind the generation of the environmental crisis' (2017, pp 14–15).

Natasha: It is vital to discuss how whiteness and privilege work, to understand how systems of oppression play out in climate justice movement spaces. There are a range of barriers within activism, particularly for young people of colour; parental support, cultural backgrounds and identity can hinder young people from getting involved or furthering their involvement as an activist (see Talukdar, 2022). Not only are there barriers to get and stay initially involved; within the movement there are further barriers of cultural insensitivity and a sense of isolation for students with marginalized identities. These barriers are forged through uneven geographies and solidarities and what we are calling 'white audacity' in movement spaces. But we also want to draw attention to how young people of colour have challenged some of the logics and workings of white privilege in these spaces over the last few years, taking their own audacious actions to create change.

Uneven geographies and solidarities

Eve: It has previously been pointed out that the global school climate strikes were 'heavily concentrated in the urban Global North' and that the climate strikes were undertaken by a 'minority of young people whose connection to international knowledge networks maps onto uneven global distributions of resources and power' (Walker, 2020, p 2). Analysing the #FridaysforFuture map, Catherine Walker notes the concentration of strikes in metropolitan cities in the Global North and the 'highly uneven' access to strikes for young people beyond metropolitan cities, as well as the 'luxury of being able to miss school, or even of being in school' for those able to strike (2020, p 2). Uneven solidarities, however, exist not only between the Global North and South, and between metropolitan and regional/remote areas. Natasha and Dani's experiences extend Walker's (2020) analysis of the 'uneven solidarity' to demonstrate the uneven conditions and opportunities among young people to engage in youth climate activism within one city – the city of Sydney, NSW, Australia.

Natasha and Dani: As the Australian Youth Climate Coalition's *Western Sydney Landscape Analysis* explains, Western Sydney is 'home to over 2.5 million people, from different cultural, socio-economic, religious and ethnic backgrounds' (Vegesana and Traurig, 2022, p 6). Western Sydney also experiences climate impacts with more severe intensity than the Eastern, Northern and Southern regions of the city of Sydney. Western Sydney has already experienced the challenges of years of 'droughts, floods, bushfires, heatwaves and the Urban Heat Island Effect' (Vegesana and Traurig, 2022, p 25). Stories of these impacts desperately needed to be recognized. Spatial inequalities and the dynamics of whiteness within SS4C have reflected representational inequalities on a societal level; in student-led climate actions held in 2018 and 2019, these voices and experiences largely went unheard and white narratives continued to hog the spotlight.[4]

In Sydney, the vast majority of SS4C organizers are from inner west and inner-city suburbs such as Leichhardt, Birchgrove, Marrickville and Newtown: suburbs close to the CBD.[5] As the Sapna report (Talukdar, 2022) details, it became increasingly difficult for students from Western Sydney to stay involved in the 2018–2019 SS4C movement, since most events/meetings were held in the city or inner west of Sydney, which can be difficult and time-consuming to get to via public transport from Western Sydney. Students from Western Sydney became used to being late and leaving early, having to fit in the extensive travel time it would take, all the while receiving no recognition for this struggle from their geographically privileged organizer peers. This extensive travelling made it more difficult for Western Sydney students to establish themselves as organizers and grow a network. Thus, students who were in more privileged positions were able to expand their own personal networks and personally gain more from being an organizer with limited accessibility issues. This geographical distance meant that Western Sydney students were often more socially isolated than students from the inner west who all lived near each other, knew each other's schools and were often in overlapping social circles. This isolation made it difficult for many students from Western Sydney to build relationships and connections in the Sydney activist sphere, limiting their capacity to step up into leadership positions, and effectively cutting their activist journey short.

Another barrier for Western Sydney students, particularly for those coming from culturally diverse backgrounds, were the cultural values of parental guardians. I [Natasha] personally struggled to gain permission from my parents to attend events and was told it would detract from my studies. It is common for migrant parents to ensure that academics are the utmost priority for their children, and thus it can be more difficult for children of migrants to get involved in such a nonstructured and radical movement [at the time] like SS4C. From my experience, my parents were extremely hesitant to let me get involved with activism, and it took a large amount of rebellion on my part for them to allow me to continue to organize strikes. I was lucky to be able to have eventually won my parents over, but for many other young people this would not be a possibility at all.

This explanation of the barriers for Western Sydney organizers and young organizers of colour doesn't mean that they cannot emerge into powerful leaders (as we do have a strong cohort of young Western Sydney activists). It is just much harder for young people, coming from those backgrounds, to be seen and represented. Activists such as us have had to overcome a lot of barriers that perhaps our white, inner-city counterparts did not have to face.

Dani: The exclusive and hierarchical relationships between inner city, predominantly white,[6] organizers and organizers of colour, largely from Greater Western Sydney, are a broader reflection of the racial and spatial exclusion that communities of colour experience in the Australian political landscape. In this sense, the organizing and relationship dynamics within the urban SS4C movement act as a microcosm of broader Australian society, where migrant communities are removed from political conversations and decisions, and when they do speak out are made to feel that those in power are not actually listening. Even when debates about migrant communities do take place, their voices are rarely the largest voice in the conversation. Minoritized members of colour frequently feel the need to take on the ways of speaking, dressing, acting and relating of white norms of communicating for their voices to be intelligible in political life; they may thus be part of the reproduction of white privilege in broader society.

White audacity in entering spaces

Natasha: Despite the social justice, 'woke' and progressive image the climate movement likes to represent, we cannot ignore how white supremacy dictates power dynamics and organizing relations. The youth-led climate movement is shaped by white logics and a white sense of entitlement. This isn't to blame the climate movement or individual white activists; this oppressive system is difficult to escape from and unlearn.

Before joining activist spaces, young people grow up and are conditioned by their different experiences of political engagement; these early experiences shape how comfortable (or not) they feel in activist circles. In my role as a peer mentor to newer student organizers, observing students who were newer to the space, white students seemed to find it easy to contribute and speak. Students identifying as Black, Indigenous and People of Colour [BIPOC], comparatively, seemed to me to be more reserved and hesitant, and less confident to step up to speak out their personal opinions and share their lived experiences. For students of colour, particularly those who are first-generation migrants (like me), the idea of us feeling represented and included in mainstream society growing up is unfamiliar. Personally, in the primary school I went to, I was one of the very few ethnically diverse children. I never felt like I belonged and felt unwanted and 'ugly' because of my physical differences from everyone else. Growing up I felt this constant discomfort, yet my inability to do anything about it made me feel hyper-aware and precautious when I entered white spaces.

I have worked to question and undo my sense of unease, which resonates with the shared experience of many other second-generation migrant young people in Australia. To my white counterparts in organizing, growing up they had felt like they belonged in Australian society: everyone looked like them. They have not experienced racial discrimination due to the colour of their skin, and thus feel comfortable entering new spaces. Young people of colour, however, have been conditioned to be quiet, step back and feel unwelcomed, whereas white students have been conditioned to speak out, take opportunities and appoint themselves into informal positions of power. It tended to be white students who would request structural changes that were hierarchical and centralised, despite the movement's intent to be nonhierarchical and remain grassroots-led. I use the term 'white audacity' to describe this confidence and boldness of white students, who are frequently unaware of the sense of entitlement they hold simply because of how they were conditioned growing up.

Eve: Natasha, the term 'white audacity' reminds me of Goenpul scholar Aileen Moreton-Robinson's analysis of 'white possessive' logics (Moreton-Robinson, 2015); beginning with the theft of the lands now known as

Australia by British colonizers from First Nations peoples with the logic of *terra nullius*. These logics continue to operate in overt and also subtle ways to maintain white dominance. According to Moreton-Robinson, white possessive logics circulate through 'commonsense knowledge, decision making, and socially produced conventions' (2015, p xii).

Natasha: The consequences of white audacity – which could be seen as a form of white possessive logics – were exemplified within the SS4C movement from 2019 to 2020 through experiences of the power dynamics surrounding media opportunities. It was common for new white students to easily feel comfortable putting their hand up for speaking/media opportunities, throwing themselves at media opportunities that were announced. This behaviour demonstrated their complete awareness of their position of privilege. Of course, the excitement of taking a 'cool' opportunity that wasn't available to other students seemed to be the main appeal; there was a lack of consideration of whether there were frontline voices who should be prioritised: 'could this opportunity be given to someone whose story should be amplified?' It took constant reminders from organizers of colour to make it a priority that BIPOC students be prioritized for media opportunities because of their lack of representation in all other mainstream areas of society. Even after this, there was still a sense of tokenism, with white students appearing to lack the understanding of why it is important to amplify BIPOC voices. Despite this change in media strategy, micro-aggressions and tensions were still present. Although on the surface SS4C was ticking all the boxes, certain racial dynamics became almost unavoidable.

In my earlier experiences in SS4C in 2019–2020, micro-aggressions and subtle acts of racism, alongside disregard of cultural and religious differences, were also common within the movement and would only be addressed if the action was publicly called out. For example, on multiple occasions, important meetings were organized on Eid, a significant date for Muslim students, making these important meetings inaccessible for students who came from Muslim backgrounds. This ignorance was called out by young organizers of colour, who carried the emotional load of noticing and pointing out that significant dates in Western cultures would not go overlooked. Further, even though young activists of colour frequently had direct place-based and family connections to climate impacts, being able to 'trace environmental impacts across two or more generations across the Global North and South' (Talukdar, 2022, p 28), this lived experience was frequently dismissed.

Eve: Natasha, your account makes me think of Cortland Gilliam's pointed question: 'How is it that white and Western climate activists come to be the faces of the global youth climate movement?' (2021, p 262). Your experience suggests that the white supremacy evident in mainstream media representations (favouring white girls in media images) also subtly played out in the earlier internal dynamics of who got the media opportunities – again, a (perhaps unconscious) sense of entitlement and 'possessive' logic about which bodies should represent other youth climate justice activists.

Natasha: There have been questions and observations of whether the youth climate movement participants are 'mainly white, middle-class and privileged' (Pickard et al, 2020, p 273). I think, in reality, there is far more diversity in the Sydney youth climate space, but the way the movement was represented in the period 2019–2020 – both through the internal distribution of media opportunities and in mainstream media practices – has perpetuated this image. The experiences of marginalized communities were overlooked, dismissed, or extracted from when a 'face of colour' was required at speeches and rallies. The meaningful learning and work that needed to be done by white organizers rarely occurred.

Audacity contra whiteness

> [W]hile a toxic ethos and acts of impunity sanctioned by institutional norms that privilege whiteness produces dehumanisation, trauma and death, bodies of colour fail to disappear. Rather than assume the position of passive subjects, they sail against the flow of whiteness, call out unjust acts and stir trouble in their struggle for justice. Through risk, adventure as well as audacious performances they emerge as ephemeral bubbles of energy that challenge national and planetary cultures of being and belonging when they stray from the well-worn path. (Lobo, 2021, p 33)

Eve: Gloria Walton, CEO of the US-based climate justice nonprofit organization The Solutions Project, writes that it is 'audacious – and requires tenacity – to have a vision for a world that cannot see' (2022, p xiii). Michele Lobo also writes about 'audacious performances' that 'challenge national and planetary cultures of being and belonging' (Lobo, 2021, p 33). Have you experienced changes to the practices in youth climate organizing spaces, perhaps because of the 'audacious' visions and performances of young organizers of colour?

Natasha and Dani: The labour of young organizers of colour, challenging the logics and practices of white privilege, has created concrete changes in the SS4C movement. In 2020, two caucuses within SS4C were created: a BIPOC caucus and a white caucus, with the white caucus set up to educate white school strikers about climate justice, intersectionality, whiteness and racialization. In early to mid-2020, the white caucus read and reflected on Layla Saad's (2020) book *Me and White Supremacy*. There have been organizational moves to change practices in media opportunities: for example, after an intervention in 2020, it became a protocol that BIPOC and regional and rural students, and students with lived experiences be prioritized for media opportunities if there were multiple students interested in the opportunity.

In September 2020, eight young people, represented by Equity Generation Lawyers, assisted by 86-year-old litigation guardian Sister Brigid Arthur, brought a class action against the [then] Federal Minister for the Environment, alleging that the Minister would breach her duty of care to young people if she approved a proposed coal mine extension project [Vickery Extension Project] in the state of New South Wales, because of the carbon that the coal would emit (Equity Generation Lawyers, n.d.). The lead litigant of this class action, and media spokesperson for this class action, was Anjali Sharma, a Melbourne high school student born in Delhi with lived experience of the impacts of climate change.[7] Changes in SS4C's organizing practices would not have been possible without the strength and persistence of young organizers of colour who have faced an uphill battle and have carried the burden of educating their white peers.

Eve: There is so much work that still needs to be done by white settlers who identify as concerned about climate justice (and I include myself here). The move in SS4C towards educating white settler strikers makes me think about Arathi Sriprakash, Sophie Rudolph and Jessica Gerrard's (Sriprakash et al, 2022, p 88) discussion of 'divesting from whiteness' through 'active, intentional and structural responses' that do not expect 'redemption' (p 87). For white settler activists, reflecting on white privilege can risk becoming an exercise in navel-gazing, narcissistic confessions of guilt and fragility (Gorski and Erakat, 2019; Sriprakash et al, 2022, p 88). Shireen Roshanravan describes, in contrast, a 'praxis of deep coalition-building' (2018, p 151). This praxis involves 'learning to see oneself in ways that are not always pleasing or easy to take' (2018, p 153): confronting one's participation in the pillars of the Master's House. This praxis can generate a sense of 'dissonance' between one's 'espoused politics' and the realization that one is 'complicit in the oppression of others' (Roshanravan, 2018, p 152).

When dissonance is actively encountered, there becomes the possibility of becoming conscious of the 'disjunctures between what is declared and what is done in the name of social justice' (Choudry, 2015, p 16). Prioritizing 'movement goals' over the need for 'recognition and validation' (Gorski and Erakat, 2019, p 804) might involve forming 'habits of accountability' (Roshanravan, 2017, p 160) to those who have been harmed by ostensibly liberatory politics both historically and in the present. It might also involve white activists stepping up to correct each other 'so that activists of colour to not need to expend energy doing so' (Gorski and Erakat, 2019, p 804).

Final thoughts

Eve, Natasha and Dani: In reflecting on these experiences together, we have wondered if the issue is with the organizing model structuring many youth-led climate justice networks, which are thoroughly enmeshed with systems of oppression. When networks like SS4C foreground 'youth', 'youth' becomes an identity marker that papers over intersectional differences; foregrounding 'youth' can default back to a universalized young person (white middle-class girl) and turn attention from building a bold intersectionality-focused climate justice movement. The School Strike organizing strategy – mobilizing mass actions and maximizing media opportunities – inadvertently has created mini-celebrities.[8]

When campaign strategy focuses on getting media attention, who gets to speak and who doesn't get to speak becomes a point of struggle, heightening the competitiveness and space-claiming practices that splinter people from each other rather than strengthen solidarity. We wonder what an alternative organizing model and structure might look like when, as Natasha has put it, "these are the only models that I know". We acknowledge the challenges of resourcing nonmainstream climate justice work, and the need for further mentorship and leadership. Further work needs to be done to laterally compare different organizing models; we gesture towards this work in this concluding section.

Responding to these accounts of uneven experiences of youth climate organizing is not about 'bringing in' excluded others into climate-organizing spaces (which fails to unsettle the 'centre'), but rather is about questioning organizing structures themselves, decolonizing youth climate justice spaces and creating alternative spaces. Thinking of 'the Master's House' pedagogical framework for teaching climate justice, we turn to Audre Lorde's speech as part of a panel at a feminist conference, to an audience of mainly white, politically progressive feminist academics: 'The Master's tools will never dismantle the master's house' (Lorde, 1984). Lorde's demonstration that social oppression cannot be solved with the tools of systems of oppression resonates with the experiences shared in this chapter: the white supremacist, capitalist and patriarchal logics that fuel the climate crisis cannot be dismantled with the logics and tools of these pillars of the Master's house. Lorde calls for attention to 'the creative function of difference in our lives' and for those 'who stand outside the circle' to 'take our differences and make them strengths' beyond the Master's House (1984, p 112).

We acknowledge the labour of those who have come before us and those who continue to lead the way in creating spaces 'beyond the Master's House'. We look laterally to other networks and witness how difference is activated as strength and power in the work of Seed Mob, Pacific Climate Warriors and Sapna South Asian Climate Solidarity. These networks emphasize collectivity and are structured by robust processes of intergenerational mentorship. In 'We are Seed Mob', the (unnamed) narrator declares:

> We are the first scientists, the oldest continuing culture. We've defended Mother Earth for thousands of generations. We have survived every shit storm that's been thrown at us. We are First Nations. We are trailblazing the path out of this crisis. We know what to do. It's time for you to follow our lead and together, we can build a future worth fighting for ... We are young, black, deadly warriors, and together, we're unstoppable. (Seed Indigenous Youth Climate Network, 2021)

Pacific Climate Warriors continually speak back to a deficit narrative of 'sinking' islands, declaring: 'We are not drowning. We are fighting!' (Tiumalu, 2015; SPREP, 2021). Through giving 'a central platform' to South Asian communities' 'stories, struggles and visions', Sapna 'aims to foster a radical solidarity in global climate justice activism' (Sapna South Asian Climate Solidarity, n.d., para 2). What makes these diverse networks so powerful is their community-based, collective and intergenerational ways of organizing and relating, beyond individualistic white Western habits of being and relating. Everyone's collective effort is needed in this ongoing struggle for climate justice.

Notes

[1] Eve Mayes acknowledges Laura Bedford for drawing the attention of the Earth Unbound collective to this open letter in 2020.

[2] We acknowledge Sophie Chiew and Netta Maiava, who are also young organizers and members of this research team, Ruchira Talukdar (project mentor) and Rachel Finneran (data manager). Sophie and Netta are leading other research publications that are currently in preparation. Our authorship practices have been informed by Max Liboiron and the CLEAR Lab's practices detailed in *Equity in Author Order: A Feminist Laboratory's Approach* (Liboiron et al, 2017). We acknowledge Sophie and Netta's contributions in research conversations during research team meetings talking through the issues discussed in this chapter. We thank Ruchira Talukdar for reading the penultimate version of this chapter and her encouraging and constructive feedback.

[3] In preparing this chapter, we have tried to find out the 'author' (individual and/or organizational) of this framework, which is widely used in climate activist circles, so that we can credit and cite them. We thank Grace Vegesana for generously sharing her knowledge of the source of The Master's House framework. At this stage, we have not been able to find the first use of this framework. This investigation suggests to us that knowledge production and dissemination in activist circles operates with different citational logics compared to those of the academy; learning resources are often built

upon and extended for different contexts and circumstances. We maintain that it is vitally important to credit and thank people for their work rather than stealing and appropriating knowledge and resources.
4   These are also tensions of scale – how to tell the stories of planetary climate change beyond universalizing science, without forsaking local struggles (Simpson and Choy, 2023, p 2).
5   Dani: There are further questions to be asked around the overrepresentation of city-based organizers within the grassroots environment organizing space, and the potential implications for worsening the city-regional divide and leaving behind regional communities. However, the scope of this chapter focuses on the experiences of organizers within urban centres.
6   When we use the term 'white organizers' and 'white students', 'white' encompasses and exceeds skin pigmentation; those visibly identifiable as white frequently have access to material and symbolic benefits, though these privileges are also shaped by other intersectional axes of privilege and oppression (including class, gender, sexuality and dis/ability).
7   However, when young people of colour are placed in the media spotlight, there is also the risk of sexist and racist abuse, as Anjali Sharma has publicly discussed (Readfearn, 2022).
8   We acknowledge and thank Ruchira Talukdar for her suggestion that we think through the limitations of the organizing model that structures SS4C.

References

Ahmed, S. (2004) *The Cultural Politics of Emotion*, Edinburgh: Edinburgh University Press.

Bevington, D. and Dixon, C. (2005) 'Movement-relevant theory: rethinking social movement scholarship and activism', *Social Movement Studies*, 4(3): 185–208.

Bowman, B., Kishinani, P., Pickard, S. and Smith, M. (2023) '"Radical kindness": the transformation of democracy by young people and environmental activism', *Council of Europe*, Available from: https://pjp-eu.coe.int/en/web/youth-partnership/knowledge-books [Accessed 17 July 2023].

Choudry, A. (2015) *Learning Activism: The Intellectual Life of Contemporary Social Movements*, Toronto: University of Toronto Press.

Colours of Resistance (n.d.) 'Colours of Resistance Archive', Available from: https://www.coloursofresistance.org/ [Accessed 3 February 2023].

Escobar, A. (2022) 'Foreword', in L. Monticelli (ed) *The Future Is Now: An Introduction to Prefigurative Politics*, Bristol: Bristol University Press, pp xxii–xxx.

Equity Generation Lawyers (n.d.) 'Sharma v Minister for Environment' [online], *Equity Generation Lawyers*, Available from: https://equitygenerationlawyers.com/cases/sharma-v-minister-for-environment/ [Accessed 15 May 2023].

Fine, M. (2018) *Just Research in Contentious Times: Widening the Methodological Imagination*, New York: Teachers College Press.

Fine, M. and Barreras, R. (2001) 'To be of use', *Analyses of Social Issues and Public Policy*, 1(1): 175–182.

Foley, G. (2000) 'Whiteness and Blackness in the Koori struggle for self-determination: strategic considerations in the struggle for social justice for Indigenous people', *Just Policy: A Journal of Australian Social Policy*, 19–20: 74–88.

Ford, A. and Norgaard, K.M. (2020) 'Whose everyday climate cultures? Environmental subjectivities and invisibility in climate change discourse', *Climatic Change*, 163(1): 43–62.

Gilliam, C. (2021) 'White, green futures', *Ethics and Education*, 16(2): 262–275.

Gorski, P.C. and Erakat, N. 2019 'Racism, whiteness, and burnout in antiracism movements: how white racial justice activists elevate burnout in racial justice activists of colour in the United States', *Ethnicities*, 19(5): 784–808.

Hage, G. (1998) *White Nation: Fantasies of White Supremacy in a Multicultural Society*, Sydney: Pluto Press.

Hage, G. (2017) *Is Racism an Environmental Threat?* London: Polity Press.

Hancock, A.-M. (2016) *Intersectionality: An Intellectual History*, Oxford: Oxford University Press.

Jafry, T., Mikulewicz, M. and Helwig, K. (2018) 'Introduction: justice in the era of climate change', in T. Jafry (ed) *Routledge Handbook of Climate Justice*, New York: Routledge, pp 1–9.

Klein, N. (2014) 'Why #BlackLivesMatter should transform the climate debate', *The Nation* [online], 12 December, Available from: https://www.thenation.com/article/archive/what-does-blacklivesmatter-have-do-climate-change/ [Accessed 3 February 2023].

Land, C. (2015) *Decolonizing Solidarity: Dilemmas and Directions for Supporters of Indigenous Struggles*, London: Zed Books.

Liboiron, M., Ammendolia, J., Winsor, K., Zahara, A., Bradshaw, H. and Melvin, J. (2017) 'Equity in author order: a feminist laboratory's approach', *Catalyst: Feminism, Theory, Technoscience*, 3(2): 1–17.

Lobo, M. (2010) 'Negotiating emotions, rethinking otherness in suburban Melbourne', *Gender, Place and Culture*, 17(1): 99–114.

Lobo, M. (2021) 'Straying beyond the well-worn path: fighting for racial justice and planetary justice', *Journal of Intercultural Studies*, 42(1): 33–45.

Lorde, A. (1984) *Sister/ Outsider: Essays and Speeches*, Berkeley, CA: Crossing Press.

McKenzie, M. (2009) 'Scholarship as intervention: critique, collaboration and the research imagination', *Environmental Education Research*, 15(2): 217–226.

Moreton-Robinson, A. ([2000] 2020) *Talkin' up to the White Woman: Indigenous Women and Feminism*, 20th anniversary edn, St Lucia, QLD: University of Queensland Press.

Moreton-Robinson, A. (2015) *The White Possessive: Property, Power, and Indigenous Sovereignty*, Minneapolis: University of Minnesota Press.

Nairn, K. (2019) 'Learning from young people engaged in climate activism: the potential of collectivizing despair and hope', *YOUNG*, 27(5): 435–450.

Pickard, S. (2022) 'Young environmental activists and Do-It-Ourselves (DIO) politics: collective engagement, generational agency, efficacy, belonging and hope', *Journal of Youth Studies*, 25(6), 730–750.

Pickard, S., Bowman, B. and Arya, D. (2020) '"We are radical in our kindness": the political socialisation, motivations, demands and protest actions of young environmental activists in Britain', *Youth and Globalization*, 2(2): 251–280.

Readfearn, G. (2022) 'Teen climate activist subjected to sexist and racist abuse amid federal court climate case', *The Guardian* [online], 12 April, Available from: https://www.theguardian.com/law/2022/apr/12/teen-climate-activist-subjected-to-sexist-and-racist-abuse-amid-federal-court-climate-case [Accessed 15 July 2023].

Roshanravan, S. (2018) 'Self-reflection and the coalitional praxis of (dis)integration', *New Political Science*, 40(1): 151–164.

Saad, L. (2020) *Me and White Supremacy: How to Recognise Your Privilege, Combat Racism and Change the World*, London: Quercus Books.

Sapna South Asian Climate Solidarity (n.d.) 'What we do' [online], Available from: https://www.sapnasolidarity.org/what_we_do [Accessed 3 February 2023].

Seed Indigenous Youth Climate Network (2021) 'We are Seed Mob' [video], Available from: https://www.youtube.com/watch?v=QEIP_qm96zw [Accessed 15 July 2023].

Simpson, M. and Pizarro Choy, A. (2023) 'Building decolonial climate justice movements: four tensions', *Dialogues in Human Geography*, DOI: 10.1177/20438206231174629.

SPREP (2021) 'World leaders told – "we are not drowning, we are fighting"', Available from: https://www.sprep.org/news/world-leaders-told-we-are-not-drowning-we-are-fighting [Accessed 17 July 2023].

Sriprakash, A., Rudolph, S. and Gerrard, J. (2022) *Learning Whiteness: Education and the Settler Colonial State*, London: Pluto Press.

School Strike 4 Climate (SS4C) (2021) '#FUNDOURFUTURENOTGAS PLEDGE', Available from: https://www.schoolstrike4climate.com/pledge [Accessed 3 February 2023].

Sultana, F. (2021) 'Critical climate justice', *Geographical Journal*, 188(1): 118–124.

Talukdar, R. (2022) 'Why North-South intersectionality matters for climate justice: perspectives of South Asian Australian youth climate activists', Available from: https://commonslibrary.org/why-north-south-intersectionality-matters-in-climate-justice/ [Accessed 3 February 2023].

Tiumalu, K. (2015) 'Koreti Tiumalu from 350.org speaks at Progress 2015', *Commons Library*, Available from: https://commonslibrary.org/koreti-tiumalu-from-350-org-speaks-at-progress-2015/ [Accessed 18 July 2023].

Vegesana, G. and Traurig, A. (2022) 'Western Sydney Landscape Analysis', *AYCC Western Sydney*. Available from: https://drive.google.com/file/d/1v7r35tEIIFLJTQdFN4PB26VNyJtNMhK-/view [Accessed 3 February 2023].

Villafaña, D., Mayes, E. and Abhayawickrama, N. (in press) '"Where are all the boys?": girl activists and climate justice activism', *Redress: Journal of the Association of Women Educators*, [online].

Vincent, E. and Neale, T. (eds) (2016) *Unstable Relations: Indigenous People and Environmentalism in Contemporary Australia*, Perth, WA: University of Western Australia Press.

Walker, C. (2020) 'Uneven solidarity: the school strikes for climate in global and intergenerational perspective', *Sustainable Earth*, 3(1): 5–17.

Walton, G. (2022) 'Foreword', in L. Thomas *The Intersectional Environmentalist: How to Dismantle Systems of Oppression to Protect People and Planet*, London: Souvenir Press.

Whyte, K. (2020) 'Too late for indigenous climate justice: ecological and relational tipping points', *WIREs Climate Change*, 11(1): 1–7.

Wretched of the Earth Collective (2019) 'An open letter to Extinction Rebellion', *Red Pepper* [online], 3 May, Available from: https://www.redpepper.org.uk/an-open-letter-to-extinction-rebellion/ [Accessed 3 February 2023].

INTERSTICE 4

Soil Geopolitics and Research as Ecological Praxis

Robin Bellingham

The soils of the place of very fast suburban growth where my three-year-old grandson lives contain waste material, including crushed rock, brick, tile, plastic pipe, concrete, steel reinforcement, asphalt, coke and slag, groundwater pollution at unsafe levels, the human-produced chemicals benzo(a)pyrene and carcinogenic polycyclic aromatic hydrocarbons, the likelihood of empty small arms shell casings and fragments of bonded asbestos cement (Strudwick, 2014). These are the effluent, residue and byproducts of commercial and suburban development and a former air base, among other things. Prevalent across the State of Victoria and especially on Dja Dja Wurrung Country where I live,[1] landforms are, among other things, covered by and made up of layers of sludge containing a range of chemicals including arsenic, mercury and cyanide that were used in the extraction of alluvial gold during the gold rush (Nicolson and Ayers, 2020).

Education and initial teacher education, the fields in which I work, have for some time reinforced ontological determinism and separation of people from the places, ecologies and communities in which they live. This separation has helped to manifest a concept of 'good education' that does not support educators to believe that it is within their remit to respond affectively, emotionally or politically to place, or to knowledge, or to support learners to do so. As Britzman (2000) noted over 20 years ago, this separation ultimately co-constitutes a void of responsibility for education institutions with regard to the most pressing and concerning issues of our time.

In this interstice I consider how an understanding of activism, research and pedagogy as forms of ecological praxis might do something to address this void of responsibility. I consider whether and how this understanding could facilitate deeper engagement of humans with ecological ways of being that are inclusive of, and which recognize, critique and respond to the relational and political, and in so doing generate different possibilities for change. To do this, I engage with soil as a model of ecological praxis par excellence. I aim not to speak for soil, but to attend to its ways of being and to engage with it on some of its own terms, and to consider some of the many ways that we can learn from closer soil relations. I conclude this interstice with some thoughts oriented towards the key aims of this book, about the implications of soil relations and the notion of ecological praxis for thinking about what planetary justice might mean.

Profane soils: energizing humble, radical knowledges and ways of being

As a kid on the Kapiti Coast of Aotearoa New Zealand, I used our black metal coal bucket as a potion cauldron and would crouch over it with a stick of kindling, dunking in miniature acacia leaves, green or yellow (decaying) karaka berries, the kernels of which I was informed were poisonous to humans and dogs, splinters of macrocarpa from around the wood-chopping block and clods of earth dug up with my fingers from under the layer of wood shards. Then in a particular shady corner, I would dole it out with a twig in small globs into the hollows of the acacia pods as an experimental buffet for tiny garden beings. The recollection of this ritual helps me to pursue again the affective maelstrom with which I was preoccupied and which most other children also understand: a deep engrossment with/in ecology and its everyday dangers, alchemy and transmutations.

The research enactments in Part III of this volume have a similar vibe to the potion cauldron, of humble concoctions, rich muck and unruly microbial interactions. The authors recognize that we are part of and inseparable from ecologies and from Country, and that it is essential that we learn from them. Also, that socioecological toxicity, imbalance and conflict are now our conditions for living and that it is our collective challenge to learn how to live with/in our ecologies with accountability and without giving up on them. But rather than pronouncing authoritative, singular truths about how we should go about this, they work with an absence of hubris, with ways of being developed over generations and with respectful grassroots experiments in living, learning and transforming with their own localized ecologies. The chapters challenge, for example, divisions between those who can and cannot determine what constitutes appropriate action and learning about ecological matters. They resist boundaries that distance knowledge from being and relationships, and that then further split up and prescribe different disciplinary kinds of knowledge such as 'education' and 'environmental science'. They jam some of the prominent narratives about climate change and sustainability that serve the interests of corporate and colonial power, and refuse to let these interests set the terms of the conversations.

These projects seem to me to enact research as a kind of ecological praxis. In other words, these examples of research, activism, resistance and solidarity are premised in lived, active participation in ecological communities. Ecology here is not just a body of science knowledge, but is also multidimensional across disciplinary and other contexts and is a form of being, inquiry, thinking and practice (Papadopoulos, 2021). Ecologies are 'the interdependent interaction between multiple forms of life' (Puig de la Bellacasa, 2021, p 198) and 'involve … the everyday experience of rootedness and belonging in our surroundings: the embodied understanding of worldly connections between different beings and environments' (Papadopoulos, 2021, p 37). Ecology is a relational way of being and doing rather than a subject. And environmental activism and ecosocial movements and research are themselves constitutive parts of ecologies (Papadopoulos, 2021), their relations, possibilities and constraints.

I consider the ecological praxis suggested in the stories of Part III of this volume alongside the praxis of soil. As my potion cauldron and other similar experiments have taught me, engaging with earth, rock, plant and water substances can connect us with and teach about the relational spirit that is in everything. As Australian Indigenous philosophies also tell us, closer attention to these kinds of relational experiences is necessary in supporting us to make essential shifts from modernity thinking and ontoepistemology, towards a more relational and ethical mode of being and learning. Soils, like the philosophies and activism collectives in these stories, demonstrate the relational nature of being quite distinctly. Their myriad organisms enact, and are media for, relations. In this way, I am considering soil as a decolonizing element that refuses traditional science thinking, which insists on identifying the basic, irreducible parts that make up more complex entities, where these basic components do not change as they compose the larger, more complex entities (Hayden, 2021). Soil relations build up compounds of matter and break them down and pass them on, remaking components that are available again to reuse as nutrients and energy. From soil communities we can notice how activism, research and pedagogical communities/ecologies are similarly not composed of discrete basic elements (things), but can instead be thought of as relations that co-produce provisional forms of energy.

Soils show us that energies produced in communities do not represent the energies of a collection of individuals. What takes place is a relation involving breaking down and building up, which effects the expression of new characteristics, as in chemistry when certain elements 'combine to form a new body possessing properties quite different from those of the bodies that have served to form it' (Le Bon, cited in Hayden, 2021, p 176), and as in many Indigenous knowledges, where it is the relation not the combination of the 'things' involved in an interaction that is significant (Yunkaporta, 2019). As in Deleuze's (1994) difference and repetition, individual entities, when part of a crowd, differ essentially from themselves, and the crowd is a new body again, possessing new potentialities that are contingent on shifting mana, vitality, lifeforce, nonindividuated relations, and intensities.

The emergence of provisional and generative energies in the research examples in Part III occurs similarly. McGregor, Christie and Kustatscher's research (Chapter 11) suggests how 'profane knowledge' emerges as a unique spirit or provisional energy that can fuel activism. The chapter reports on the ways in which young people experience external denials of their 'profane knowledge' – that is, their own lived forms of knowledge on matters that are of central and crucial relevance to them, their lives, education, futures and ecologies. In many cases, because these people are categorized as 'youth', their views and voices are dismissed by the authorities, and because their experiential knowledge reflects a response proportionate to the real context of planetary and personal existential crisis, it is deemed too radical or ideologically charged by authorities. Their research suggests that resisting disavowals of, and attending to, profane knowledge, with its own adaptative and relational power, can vitalize the community in ways that can provide an emancipatory breach from dominant knowledge systems.

I suggest that 'profane soils' can describe the ground from which profane knowledge and other profane vitalizing energies can emerge in research and activism. The profane is a generatively slippery term, implicating both the everyday, earthy and humble, and the radical, irreverent, noncompliant, nondeferential and sacrilegious. Katherine McKittrick (2006, 2015) theorizes demonic grounds as places from which similar refusals of recognized authority as the legitimate centre and basis of knowledge and transformative action can be enacted. As part of her critique of the relational and ontological logics of white modernity, McKittrick theorizes demonic grounds as the places Black women inhabit who live and think refusals of and liberations from knowledge systems that seek to own, possess, extract, have and exclude.

McKittrick (2006) notes that from demonic grounds, observation of how present modes of being function unjustly and create invisibilities becomes possible, but also that from these places it is possible to see and to be human in other radical and creative ways. As those who inhabit demonic grounds are positioned to have a clearer view of the ecological and geopolitical conditions that inscribe bodies, they are also positioned to see and create alternative lines of connection, cosmologies and stories of being (McKittrick, 2021). Profane soils might be theorized as beneficial media for similarly resistant relations, and grounds from which to reimagine categories of inclusion, exclusion, authority and margins. These soils sustain and feed wayward, radical and potent stories, knowledge, agency and ways of being, activism and research.

Soil geopolitics

To avoid romanticizing or working reductively with soil knowledge – for example, by elaborating wishful metaphors or by superficially grafting soil-like processes onto existing dominant frameworks of understanding – we need to have an affinity for how soil itself expresses a different mode of being. Soil expresses the enmeshment of all things – for example, that the geological and the political are inseparable. Chemical toxins are now an enduring part of the relational continuities connecting us with plants, animals, earth, air and water. Human-made chemical toxicity is so deeply embedded in and circulating through soils, machines, human practices, industry and throughout ecologies. The question is now not how to process and eliminate this through rebalancing natural cycles, but how to sustain life in permanently contaminated ecologies (Papadopoulos, 2021). For this reason, a model of ecological praxis implicates deeper and more creative geopolitical understandings and action, resistance and solidarity.

The contaminated soil and sludge described in the opening of this interstice expresses the geopolitics of soil life in a particular place, including soil's entanglement with colonial capitalism and the marks and inscriptions this leaves. Embedded in these contaminated sediment layers are the ghosts of those entities damaged, subjugated and destroyed as resources and byproducts of colonialism and modernity. Hauntings can potentially catalyse clearer visions of the stark realities of the present, past and future (Bubant, 2017), and these sedimentary and toxic hauntings express most eloquently and accurately what the human condition and fate are: that we are 'fossil colonial animals that are engineering our own demise' (Bubant, 2017, location 2391). But hauntings do not just express past trauma and damage; they are also a demand for something to be done (Gordon, 2011).

To build capacity to recognize, pay attention to, care about and take action on these hauntings and the past/present/future of our ecologies, an ecological praxis needs to engage explicitly, routinely and experientially with the geopolitical. Historical and present (often deliberately hidden) forms of human power and hegemonic knowledge, including Western science, colonialism and capitalism, are the basis for modernity's ethics and justification of instrumentalization, accumulation, and abuse of the planet. Their general invisibility is itself a means of ongoing violence and a major obstacle to growing our capacity for ethical responses in these relations. Science, in particular, continues to be presented and understood by many as emblematic of a superior knowledge system that is value-neutral and our primary means for finding ecological solutions, while simultaneously strengthening colonizing forms of power, including notions of 'nature' as an external object and resource. The ambivalent and complicit relations of science, industry and colonial, corporate and political institutions with planetary violence need to be closely attended to in research and actions towards planetary justice.

Education and research need to engage as a matter of course with the geopolitical. For example, in teaching, learning and knowledge about soil, this means engagement with practices of soil displacement, compacting, choking, degrading and poisoning in waste disposal and landfills, resource extraction and processing, science laboratories, topsoil removal, forestry and agriculture, housing development, burials in cemeteries, and roadworks, and with the inscriptions of these on bodies in situated ecologies, including sickness, extermination, pollution, erosion, fires, flooding, barrenness, and more-than-human subjugation and exploitation. These engagements need to be approached in nonabstract ways, in local situations, in relational experiences with soils and ecologies, and with these colonial capitalist forces – for example, in the ways explored in discussions of On Country Learning (by Fricker, Chapter 9), the Sunderbans (by Mukerhjee et al, Chapter 10) and Western Sydney (by Abhayawickrama et al, Chapter 12) in this book.

A particularly potent pedagogical affordance of soil is that of all 'natural bodies', it provides one of the clearest examples of how life and degenerative processes are inseparably entangled. Entities in soil degrade and decompose each other and for each other. These processes are integral to giving away what is no longer needed, which can then be put into nourishing overall metabolism and into transformation of the ecology through the mobilization of new energies (Puig de la Bellacasa, 2021). In this way, soil challenges the privilege that is extended to processes of creation, production, building up, accumulation and excess in modernity. Modernity's unbalanced celebration of life and growth misses the essential processes of 'breakdown and circulation of matter that rebalance generation, productivity, and excess' (Puig de la Bellacasa, 2021, p 197).

Puig de la Bellacasa emphasizes the vital, contemporary question: 'How can we foster our capacity for breakdown as an elemental affinity to all in a world stunned by the pervasive chemical compounds that will not break down and which are altering elemental cycles?' (2021, p 216). Lyons (2016) suggests we think of 'decomposition as life politics'. Lyons observed how Colombian farmers on lands in the midst of drug wars worked with dying as part of the transformation of being; emergences from regenerative decay as practices lived and learned. The farmers collectively reshaped the conditions of their lives 'not by transcending these conditions, but rather by sinking into them, slowly turning them over, aerating, and breathing in new life that also potentiates different possibilities for and relations to death' (Lyons, 2016, p 65). Of the Amazonian soils themselves, Lyons remarks that they express 'weariness and recoil from the impossibility of existing under the relentless strain of extractive conditions' (2016, p 74). Their destruction is not a finite death, but marks exhaustion, a turning away from certain ways of life. Lyon's decomposition as life politics 'compels us to delinearize the cosmological tale, to turn origins into cycles, lines into spirals' and 'to complicate neat models' (Puig de la Bellacasa, 2021, p 200). It implicates the expectation of fragility, vulnerability, decline and deterioration as inherent to ecological praxis, and suggests that these are states significant to the possibilities of transformation.

The soil geopolitics that are emergent here draw attention to the increasingly clearer need for us to engage with the ways of being, including the political activity, of more-than-human others, and the implications and relationship of this political activity to the planetary toxicity, excess and imbalance we all must reckon with. An aspect of the degenerative geopolitics of ecologies which we should also learn from and take seriously is 'ecological insurgency' (Papadopoulos, 2021, p 47), which occurs when ecologies mutiny against that which attempts to manipulate and control them. The resistance of beings classed as pests to the pesticides created to target them is an example. Anthropogenic chemicals, technologies and practices introduced as means to control the environment are now widespread and their activities are unpredictable and complex, including seemingly unstoppable proliferation, viral mutation, carcinogenesis, eutrophication, global warming and endocrine disruption (Papadopoulos, 2021).

Activism as ecological praxis: what does this suggest for a concept of planetary justice?

In this interstice I have proposed that the chapters in Part III of this volume collectively demonstrate a kind of ecological praxis which I liken to earthy and unruly microbial interactions. Ecology here is more than a body of science knowledge. It is multidimensional across disciplinary and other contexts and is a form of being, relating, inquiry, thinking and practice.

I suggest that this praxis has similar pedagogical, affective and relational affordances to those we know from our own, often unsanctioned, childhood immersions in play and learning with/in backyard or local ecologies and soils. This kind of soil work and play is engagement as part of earthy, humble, multi-organism communities. Like the forms of inquiry in Part III of this volume, the transformative potential of soils is in part their demonstrative refusal of deterministic separation of their elements.

In this interstice I have co-opted McGregor et al's term 'profane knowledge' (Chapter 11) to describe the forms of knowledge and experience from which injustices become visible, and which enable humble but radical and sometimes transformative responses to these injustices. I have considered how the notion of profane soils is also generative in reflecting on the nature of the grounds of places and ecologies from which such transformative knowledge and agency might be sustained.

The degeneration that soils enact is a potent pedagogical provocation in the geopolitics of ecological praxis. As many Indigenous peoples are well aware, soil and other communities rely on states of decay, decline, insurgency and/or turning away (Povinelli, 2016) – for example, when poisoned or subject to the violences of colonial management. These changing states reflect inevitable and necessary shifts in the relations of the community, for metabolism, adaptation and transformation. Soil challenges the imperative in modernity for continuous growth and so-called improvement, demonstrating the limitations of the question of how to become 'better' all the time. Soil tells us that decay, slowing down, lying low and insurgency have essential value that is misunderstood and unrecognized in modernity. Learning about this seems necessary to help us approach the question of how to keep on in an ecologically relational world in which damage and toxicity are now enduring. Refusals to abandon imperatives for accumulation and progress, and continually strengthening efforts to manage, control and subordinate have not and cannot achieve positive outcomes or change in relation to this aim.

Among other things, turning away, slowing down, death and insurgency will have to mean, as Fricker says in this volume, being 'prepared to pay a personal cost in time, wealth or risk' (Chapter 9). These costs are gravely demanding to accept, implement and adjust to. They mean dismantling instead of preserving the Master's House (Lorde, 1983), as Abhayawickrama et al discuss (Chapter 12). For many people in developed countries, acceptance of these necessary states must also occur in relation to our own habits of being, accumulation and consumption, to white privilege and white supremacy, and to notions of white settler innocence.

Modernity demands that politics and justice are organized around questions of who should take decisions over what and on whose behalf, a way of thinking which has helped to entrench violent forms of dominance and oppression. This interstice suggests that planetary justice is not just about what to do and who should decide, as in autonomous independent subjects deciding for and acting on objects. Communities working for activism and solidarity in soil-like ways resist such fixed power dynamics and boundaries. Activism, research and pedagogy as forms of ecological praxis raise different kinds of questions about politics and justice, such as questions about the diverse ways in which communities and entities do and might organize relations and become catalysed for change, or how they live and sustain different ways of being in solidarity with Country, as Fricker discusses in this volume (Chapter 9). This means respect for and paying close attention to their everyday flows, cycles, politics and transformations, and it necessitates getting our hands dirty and paying some costs for more bodily, emotional, critical, accountable, ethical, caring relations with/in these ecologies.

Note

[1] I write this interstice on Dja Dja Wurrung Country in Victoria, Australia. I acknowledge the Traditional Djaara custodians and Elders, past, present and emerging, and acknowledge this Country as one I was not born into, but to which I am grateful to have the privilege of living and learning with. I also acknowledge the Kapiti Coast of Aotearoa New Zealand as my home and first teacher, and Maori iwi including the Rangitāne, Muaūpoko and Ngati Toa as its first peoples and custodians.

References

Britzman, D. (2000) 'Teacher education in the confusion of our times', *Journal of Teacher Education*, 51(3): 200–205.

Bubant, N. (2017) 'Haunted geologies: spirits, stones and the necropolitics of the Anthropocene', in A. Tsing, H. Swanson, E. Gan and N. Bubant (eds) *Arts of Living on a Damaged Planet*, Minneapolis: University of Minnesota Press.

Deleuze, G. (1994) *Difference and Repetition*, P. Patton (trans), New York: Columbia University Press.

Gordon, A. (2011) 'Some thoughts on haunting and futurity', *Borderlands*, 10(2): 1–21.

Hayden, C. (2021) 'Crowding the elements', in D. Papadopoulos, M. Puig de la Bellacasa and N. Myers (eds) *Reactivating Elements*, Durham, NC: Duke University Press.

Lorde, A. (1983) 'The master's tools will never dismantle the master's house', in C. Moraga and G. Anzaldua (eds) *This Bridge Called My Back: Writings by Radical Women of Color*, New York: Kitchen Table Women of Color Press, pp 94–101.

Lyons, K. (2016) 'Decomposition as life politics: soils, selva, and small farmers under the gun of the U.S.–Colombia war on drugs', *Cultural Anthropology*, 31(1), Available from: https://journal.culanth.org/index.php/ca/article/view/ca31.1.04/345 [Accessed 12 March 2024].

McKittrick, K. (2006) *Demonic Grounds: Black Women and the Cartographies of Struggle*, Minneapolis: University of Minnesota Press.

McKittrick, K. (ed) (2015) *Sylvia Wynter: On Being Human as Praxis*, Durham, NC: Duke University Press.

McKittrick, K. (2021) *Dear Science and Other Stories*, Durham, NC: Duke University Press.

Nicolson, O. and Ayers, A. (2020) 'Sludge in Victoria: what is it and why?', Press Release, *Ecology and Heritage Partners*, Available from: https://www.ehpartners.com.au/newsroom/sludge-in-victoria:-why-itand39;s-important [Accessed 12 March 2024].

Papadopoulos, D. (2021) 'Chemicals, Ecology and Reparative Justice', in D. Papadopolous, M. Puig de la Bellacasa and N. Myers (eds) *Reactivating Elements*, Durham, NC: Duke University Press, pp 34–69.

Povinelli, E. (2016) *Geontologies: A Requiem to Late liberalism*, Durham: NC: Duke University Press.

Puig de la Bellacasa, M. (2021) 'Embracing breakdown: soil ecopoethics and the ambivalences of remediation', in D. Papadopoulos, M. Puig de la Bellacasa and N. Myers (eds) *Reactivating Elements*, Durham, NC: Duke University Press, pp 196–230.

Strudwick, D. (2014) *Environmental Audit Report Stage 6A, Williams Landing Development, Palmers Road, Williams Landing, Victoria*, Melbourne: AECOM Australia.

Yunkaporta, T. (2019) *Sand Talk: How Indigenous Thinking Can Save the World*, Melbourne: The Text Publishing Company.

POSTSCRIPT

The Earth Is Undone

Alicia Flynn

The Earth is undone
Unbound by the laws of the Tao, the heavens,
The Earth unravels
The Earth unravels as people with heavy feet and heavy hearts
Plow into the skin and disembowel
The contents
Spilling the guts of the Earth for their fleeting riches
To cover the pain and torment, the void in their hearts.

In this Earth I sit behind a laptop and type
Alone, but never alone
In a room, in a home, connected-disconnected
The blue light, the bad posture,
The overstimulation of some senses and death of stimulation of bodied being
My social animal spirit shrivels
Once a plump plum now a prune
Who have I become, what have I done?

The Earth is undone bit by bit
And in turn *this* Earth undoes human animals bit by bit
Scholarship cannot survive this mercurial dislocation
This technologically advanced evisceration.

'Follow the body and the rest will follow'
Leave the Cartesian split behind
Follow the relations of the heavens and Earth
And the Earth will be bound again, but differently.

In the practice and images of reflection
We see only ourselves
We learn Narcissus against Narcissus
Not love despite and because of different differences.

Individuals are composted by the organised chaos
of thermodynamics
Complexity will only increase, uncertainty certain
Dance with the light of the moon in the feast of the heavens with the bodies of the Earth
Or become, once more, stardust, metals in mobile phones, inertia unbound.

And then …

Tuning into the entangled affections of First Nations' fire relations
I learn to love again but differently
Love not for the destiny of studies and qualifications
But love for my relational body, for the subterranean soul of soil.

I step down from the Ivory Towers
Into the garden of a primary school
A return to teaching that is a turn to otherwise
A dance of becoming-with the worldliness of children and worms, fungi and rot.

Scholarship will live in this new biome
Or not at all.

Index

References to figures and photographs appear in *italic* type; those in **bold** type refer to tables. References to endnotes show both the page number and the note number (231n3).

A

Aboriginal Cultural Heritage Act 2021 28
Aboriginal peoples
  Australian Constitution 102
  colonialism, effects of on 40–2
  environmental racism toward 43–5
  see also ancient wisdom; First Nations peoples; Indigenous peoples
accomplices 167, 168, 173
activism *see* climate activism; School Strike 4 Climate (SS4C)
adaptive capacities 182, 190
Adivasis 106–7, 108, 111, 114n4, 131, 133
Aga, Aniket 107–8, 109–10, 111
Agarwal, Anil 127
agency 11, 26, 33, 46, 169, 179, 201, 206, 208
Agozino, Biko 141–2
Aila *see* Cyclone Aila
air quality 2
allies 49, 125, 134, 166–7, 168, 173
Amin, Samir 142
Amphan *see* Cyclone Amphan
ancient wisdom
  about 25–26, 92
  climate justice and 31–3
  solutions within 34–5, 36, 234
  see also Aboriginal peoples; First Nations peoples; Indigenous peoples
Anthropocene 6, 9, 83, 179, 181
anxiety *see* climate anxiety
ANZUS Treaty 166–7
aquifers 54, 55, 58–9, **61**, 63, 64
art *see* bark painting; sand art
Arthur, Sister Brigid 225
Arya, Dena 202
audacity *see* white audacity
Australia
  anti-coal resistances in 98, 99, 101–6, 111–12, 113–14
  bushfires 1, 164, 220
  climate activism in 3, 213–14
  climate change impacts 164
  Constitution 102
  education system 169–71, 172
  life force 26, 73–4, 80, 84, 90
  Native Title Act (NTA) 102–3, 104, 106, 114n1, 114n3
  Racial Discrimination Act (RDA) 102–3
  water management 42–3, 43–5, 49
  Western Sydney 220–1
  see also Country; First Law (Law of the Land); Murray-Darling Basin (MDB); settler colonialism; Waking up the Snake metaphor

B

Bai, X. 180
bank heists 167
bark painting 74, 80, 83
Barrow, J. 34
Bassey, Nnimmo 141
Baud, M. 125
Bawaka Country 29
Bebbington, A. 130
Beetaloo Basin 2
bhatir desh see Tide Country *(bhatir desh)*
Biermann, F. 12, 180
Bird Rose, D. 34
Black, Indigenous and People of Colour (BIPOC) *see* climate activism; youth
#BlackLivesMatter 219
boodja 25, 27, 32, 33
  see also Country
bottom-up *see* justice; solidarity
Britzman, D. 232
Bulkeley, H. 11
Bundle, Yaraan Couzens 78
Bunjil (creator spirit) 74, 82, 84

INDEX

bushfires 1, 164, 220
business-as-usual, colonization as 26, 30, 31, 33, 35

C

capacities 180, 181, 182, 189, 190, 191, 192
capitalism
  'Black Summer' 1
  confronting 142
  disaster capitalism 200
  effects of 3, 9, 143, 202
  extractive 73, 97, 106, 173
  governance systems under 141
  growth, and 203–4
  revanchist movement 202–3
  solidarity against 145–8
Capitalocene 9
care, ethics of 26, 36
Carmichael coalmine 97, 98, 99–100, 105–6
cetaceans *see* whales
Chickaloon Village Traditional Council 105
Chiew, Sophie 227n2
Chinna, N. 30–1
Choudry, A. 217
Chowdhury, Chitrangada 107–8, 109–10, 111
Choy, Tim 16
Christie, Beth 197, 204–7, 235
citizenship *see* ecological citizenship
Citizens' Protection Declaration 77–8, *78*
Clark, Nigel 13
class consciousness 143, 147–8, 169
class solidarity 7, 140, 142–3, 147–9
climate activism
  BIPOC students 222, 223, 224–5
  class solidarity, as 147–9
  Indigenous peoples 3, 73, 79
  youth 34–5, 73, 146–7, 196, 206–7, 208–9, 213–14
climate anxiety 197–200, 200–2, 206, 207–8
climate care 29, 88
climate change
  impacts of 101, 155, 164, 179, 180, 183
  injustices and 8–9
  symptom, as a 147, 148
'Climate Change and Oceanic Responsibilities' 81–2, *82*
Climate Change Conferences (UN) *see* Conferences of the Parties (COPs)
climate justice
  about 10, 213–14
  ancient wisdom and 31–3
  climate care, as 29, 88
  human rights and 113
  Indigenous 98
  intersectionality of 6, 100–1, 214–16, 226
  see also climate activism
Club of Rome 203–4
coal production 2
  see also fossil fuel industry

co-creation 206
Collective, the *see* Earth Unbound Collective
collective action frames 122–3, 124–6, 128–9, 132, 134
collective consciousness 26, 28, 32–3, 34–5, 36, 171
colonialism
  business-as-usual, as 26, 30, 31, 33, 35
  education systems 169–71
  effects of 3, 9, 29–31, 40–2
  epistemic violence, as 26–8
  extractive 31
  freshwater resources, impact on 40–1
  postcolonialism 97–8, 101, 110, 111, 114
  settler 72, 97–8, 101, 103, 110, 111, 114, 144, 219
  slow violence of 14, 27, 29–31, 144–5, 152, 236–7, 239
  water 43
  see also decolonization; postcolonialism
Colours of Resistance 216
comrades 168–9, 173–4
Cone, James 167–8
Conferences of the Parties (COPs) 2, 12, 80–1
Connelly, L. 5–6
consciousness
  class 143, 147–8, 169
  collective 26, 28, 32–3, 34–5, 36, 171
  ocean, as an 92
  planetary 16
  waking up of 27, 29, 31
  white/Western 72, 83, 89
consent 105–6, 107–8, 115n9
  see also manufactured consent
contradictions of solidarity 6–7
converts 167–8, 173
COP *see* Conferences of the Parties (COPs)
coping capacities 182
co-production *see* knowledge co-production
Coulthard, Glen Sean 9, 144
Country
  about 36n1, 87, 165–6
  On Country Learning 171, 173
  guardianship responsibility for 26
  language, importance of 166–9
  narration of 31–2
  solidarity with 169, 171–4
  songlines 28–9, 73, 76–7, 79
  stewardship of 28–9, 30
  teacher, as 172–3
  Waking up the Snake metaphor 27–8, 31–2, 33, 34, 35, 36
  see also Saltwater Country
COVID-19 pandemic 2, 4, 10, 158, 184, 185
Crenshaw, Kimberlé Williams 100
criminology *see* green criminology
crocodiles 74, 83, 179

245

crosscurrents *see* decolonial crosscurrents
Cross Curriculum Priority 172
cup of tea method *(kapati)* 26
Curnow, J. 6
curriculum 171–2
　see also education
Cyclone Aila 60, 61
Cyclone Amphan 82, 155–9, 184

D

da Cunha, D. 91–2
Dakota Access Pipeline 105
dancing *see* sand art
Davison, A. 198
Death of a Discipline (Spivak) 13
decolonial crosscurrents 14, 71, 72, 83, 88
decolonization
　decolonial praxis 71–72, 83, 90
　failures of, facing 152–4
　pedagogy, of 163–4, 170–1, 171–3
　regeneration, as 26, 31, 35–6
　risk burdens of 167
　risks associated with 72
　true, requirements of 87, 142
　see also colonialism; postcolonialism
defenders 15, 140, 143–4
deference politics 143
degrowth 49, 203
Deleuze, G. 235
democracy *see* economic democracy
demonic grounds 235
descriptive approach 181
Deshia Kondh 134n2
diagnostic framing 124
　see also collective action frames
digital 83, 84
digital performances 84
disaster capitalism 200
disavowal *see* institutionalized disavowal;
　profane knowledge
displacement 43, 61, 114, 122, 133
dissonance 146, 225
dividual 91
Dixon, C. 217
Dja Dja Wurrung 163, 164, 232, 240n1
Dongria Kondh people 122, 123, 126, 128, 129–31, 132, 134, 134n1–2
Dreaming 27, 30–1, 165, 168, 173–4, 174n1
　see also Whale Dreaming (Hunter)
Dutt, Sanjana 15
duty of care 2, 225
Dyson, J. 129, 131

E

Earth Unbound Collective 1–4, 4–8, 29
ECJ *see* Education for Climate Justice (ECJ)
eco-anxiety *see* climate anxiety
ecofascism 199–200
ecofeminism 10, 213
ecological citizenship 48, 50
ecological justice
　about 10, 40, 46, 87
　capitalism, and 141
　farmers, and 48–50, 92
　planetary justice and 13
ecological praxis 16, 232–3, 236, 238–9
ecology 28, 92, 234
economic democracy 129–30
ecosocialism 10
education 130–2, 163–4, 169–71, 172, 201–4, 207–9
　see also institutionalized disavowal;
　pedagogy, decolonizing
education for a just transition 201, 207, 209
Education for Climate Justice (ECJ) 196, 197–8, 199–201, 203, 204
Eight Ways pedagogy 170–1
eminent domain 108, 114n6
Emirbayer, M. 208
emotional literacy 197
emotional reflexivity 200
emotions 197–200
　see also climate anxiety
encounters 71, 91, 92
energies *see* kinetic energies; sacred energies
environmental defenders *see* defenders
environmental justice
　about 10–11, 40, 87
　green criminology, and 39–40
　planetary justice and 13
environmental racism 43–5
　see also racism
Equity Generation Lawyers 225
Escobar, Arturo 181
ethics of care 26, 36
ethnographic research 99–100, 109, 126, 179, 185
Extinction Rebellion (activist group) 3, 216
extraction education 131
extractive capitalism 73, 97, 106, 173
extractive colonialism 31
extractivism 15, 49, 98, 101, 112–13, 123, 128, 140, 149, 207, 237–8
Eyerman, R. 125

F

failures of decolonization 152–4
farmers
　decay, working with 237
　ecological citizenship of 48–50, 92
　feminization of 59
　on floodplain harvesting 45–6
　risks faced by 44
　water markets and 42–3
First Law (Law of the Land) 25, 26, 27, 28, 35, 36, 47
First Nations peoples
　about 163

Country, concept of 165–6
Dreaming 174n1
 language, impact of on 166–9
 pedagogical sovereignty of 163–4, 170–1
 sacred energies 73
 solidarity with 169
 sustainability, concept of 172
 see also Aboriginal peoples; Indigenous peoples; *specific groups (e.g., Dja Dja Wurrung)*
floodplain harvesting 43, 45–6
forces *see* kinetic energies; sacred energies
forest fishing 184–5, 186, 187, 188–90, **189**
forest rights 98–9, 106–10, 110–11, 112, 114, 115n8
Forest Rights Act (FRA) 102, 107, 108, 109, 110–11, 114n5, 115n7–9
fossil fuel industry
 climate change, and 101
 resistance to 3, 97–9, 101–6, 111–12, 113–14
 see also Carmichael coalmine
Foster, J.B. 142, 145–6
fracking 2
frames *see* collective action frames
framework *see* Master's House framework
Freire, Paulo 146–7, 169, 171
freshwater
 aquifers 54, 55, 58–9, **61**, 63, 64
 floodplain harvesting 43, 45–6
 groundwater 55–6, 58–9, **61**, 62
 intrinsic value of 39, 40, 46–7, 49, 50, 92
 management of 41, 44, 63–6
 resources 39, 45
 rights of 46–7
 water markets 42–3, 49, 91
 see also Waking up the Snake metaphor
#FridaysforFuture 220
'Fruit Gathering' (Tagore) 159
futures 13

G

Gabrys, J. 13
Gerrard, Jessica 225
Gilliam, Cortland 223
Global Witness 140, 143–4
Gómez-Barris, Macarena 9
Gosaba 184–5, 185–8, *186*
Gram Sabhas (village councils) 107, 109, 114n5, 115n9, 128
green criminology 39–40
Greenpeace 98, 108–10, 112, 113, 115n7
ground up *see* justice; solidarity
groundwater 55–6, 58–9, **61**, 62
Gumtaj clan 74
Gunditjmara 73, 74, 76, 77–8, 79, 84
Gupta, M. 131

H

Hage, Ghassan 218–9
harmony 30–1, 32, 34, 75

Hartwig, D. 43
harvesting
 floodplain 43, 45–6
 rainwater 54, 57–8, 64–5, 66
Harvey, J. 167–8
hauntings 236
healing
 saltwater 72, *80*, 81
 sea country 80, 81
Helferty, A. 6
Hickell, J. 49
Hickey, C. 12–13
house *see* Master's House framework
human rights
 climate justice and 113
 self-determination, to 15, 97, 98–100, 101–3, 106–8, 111–12, 113–14
 water, access to 44, 56, 65
 see also rights
hyper-self-reflexivity questions 152–3

I

ILUA *see* Indigenous Land Use Agreement (ILUA)
India
 anti-coal resistances in 98, 99, 106–10, 111–12, 113–14
 Constitution 65, 66
 Cyclone Amphan 82, 155–9, 184
 Forest Rights Act (FRA) 102, 107, 108, 109, 110–11, 114n5, 115n7–9
 Supreme Court 108, 123, 126, 128, 130, 132, 134
 see also Niyamgiri Mountains; Niyamgiri Movement
Indian Sundarban Biosphere Reserve (SBR)
 about 182–4
 climate change, impact of on 179, 180, 183
 field research in 185–90
 forest fishing 184–5, 186, 187, 188–90, **189**
 mangrove forest of 72, 179
 riskscapes in 184–5
Indian Sundarban Region (ISR)
 about 55–8, *56, 57*
 freshwater crisis 54, 58–60, **61**, 61–3, 66–7, 91
 water management 63–6
Indigenous cultures 90–1
Indigenous justice 98
Indigenous Land Use Agreement (ILUA) 104, 105–6, 114n3
Indigenous methodology 26, 29
Indigenous pattern thinking *see* sand talk
Indigenous peoples
 activism by 3, 73, 79
 climate change, impact of on 101
 colonialism, effects of on 3, 29–31, 41–2
 environmental racism toward 43–5
 fossil fuel industry, resistance to 3, 97–9

land rights 98–9, 101–3, 110–11, 113, 213
life force 73–4
perspectives of 89
self-determination, right to 15, 97, 98–100, 101–3, 106–8, 111–12, 113–14
violence faced by, systemic 12, 101, 109–10, 111, 134, 145
see also Aboriginal peoples; ancient wisdom; First Nations peoples; *specific groups (e.g., Noongar)*
Indigenous Peoples and Local Communities (IPLCs) 101
Ingersoll, Karin 72, 83
injustice
of climate change 8–9
global environmental 10
water 43, 63, 66
see also justice
innovators 125, 134
institutionalized disavowal 200–1, 203, 204–5, 207–8, 235
intellectuals *see* movement intellectuals; popular intellectuals
intergenerational justice 11, 44, 199, 208, 213, 227
Intergovernmental Panel on Climate Change (IPCC) 4
intersectionality
about 100, 228n6
climate justice work, and 6, 100–1, 214–16, 226
of injustice 3, 4, 6, 215
solidarity, of 112–13, 192, 200, 215
intrinsic value 39, 40, 46–7, 49, 50, 92
ISR *see* Indian Sundarban Region (ISR)

J

Jafry, T. 213
Jagalingou *see* Wangan and Jagalingou (W&J) Traditional Owners
Jal Jeevan Mission (India) 59–60
Jana, Pranati 62
Jeevanshalas 132
Jeffrey, C. 129
Jones, C.A. 198
Joseph-Salisbury, R. 5–6
Jungara (time traveller) 32
justice
about 10, 14, 87
bottom-up 16
see also climate justice; ecological justice; environmental justice; Indigenous justice; injustice; intergenerational justice; planetary justice
just transition *see* education for a just transition
Just Transition Commission 201
Juukan Gorge 28, 36n3

K

Kalfagianni, A. 180
Kalinga Institute of Social Science (KISS) 131
Kanaka Maoli 72, 83
kapati (cup of tea method) 26
Keck, M. 182
Keystone Pipeline 105
Kimberley 26, 27, 30–1, 74, 76
kincentric ecology 28, 92
kinetic energies 14, 71–2, 74, 76, 83
Klein, Naomi 219
knowledge *see* ancient wisdom
knowledge co-production 170, 180–1, 182, 187–90
Kolkata 56, 58, 60, 63, 71, 76–7, 155–9, 183
Kondh tribe 134n2
see also Dongria Kondh people; Kutia Kondh people
Kornup (land of the setting sun) 32
Kothari, A. 130
Kovach, M. 29
Kovel, J. 147–8
Kumirmari *see* Gosaba; Indian Sundarban Biosphere Reserve (SBR)
Kutia Kondh people 122, 123, 127, 128, 134n2

L

Land 36n1
land defenders *see* defenders
land rights 98–9, 101–3, 110–11, 113, 213
language 166–9
Latour, Bruno 6
Law of the Land *see* First Law (Law of the Land)
Learning for Sustainability (LfS) 201, 204–6
legal personhood 47
life force 26, 73–4, 80, 84, 90
'Limits to Growth' (Club of Rome) 203–4
Lloyd, V. 167–8
Lorde, Audre 226
Lyons, K. 237–8

M

Mabo decision 102–3
MacKay, Kevin 146
Madarrpa clan 74
Mahan 98, 99–100, 108–10, 111–12, 113–14, 115n7
Mahan Sangharsh Samiti (MSS)/Mahan Resistance Front 108, 109, 110, 115n7
Mahatma Gandhi National Rural Employment Guarantee Act (MGNREGA) 65–6
Maiava, Netta 227n2
Majhi, Siuli 61
manufactured consent 105, 107–8, 109, 111
'Manufacturing consent' (Chowdhury and Aga) 107–8

INDEX

Maoists 132–3
markets *see* water markets
Marks, E. 199
Martuwarra 27, 30–1, 47
Martuwarra Dream 30–1
Martuwarra Fitzroy River Council 47
Marxism 142, 144–5, 148
Master's House framework 218, 219, 225, 226–7, 227n3, 239
Mawdsley, E. 111, 114
McGee, T. 168
McGlade decision 114n3
McGregor, Callum 197–204, 235, 238
McKittrick, Katherine 235
MDB *see* Murray-Darling Basin (MDB)
Me and White Supremacy (Saad) 225
media
  climate change coverage 60, **61**
  white audacity and 223–4, 225
metaphors *see* Waking up the Snake metaphor
micro-aggressions 223
Milgin, A. 28, 35
mining 122, 123
miny'tji see sacred designs *(miny'tji)*
Mische, A. 208
Mollah, Tanuja 62
Moore, Jason 9
more-than-human life force *see* life force
Moreton-Robinson, Aileen 219, 222–3
Morgan, Jonathan 8
motivational framing 124
  see also collective action frames
movement allies 125, 134
movement intellectuals 125, 134
MSS *see* Mahan Sangharsh Samiti (MSS)/Mahan Resistance Front
Mukesh (Dongria Kondh leader) 129, 130
multisensory storying 82–84
multispecies justice 10
murder 87–8, 143–4
Murray-Darling Basin (MDB)
  about 40–1
  ecological citizenship in 48–50
  floodplain harvesting 43, 45–6
  water market 42–3

N

Narmada Bacchao Andolan (NBA) 129, 132
native title 102–3, 104–5, 106, 110–11, 114n1, 114n3
Native Title Act (NTA) 102–3, 104, 106, 114n1, 114n3
NBA *see* Narmada Bacchao Andolan (NBA)
networked resilience *see* social resilience
networked solidarities 180
ngilgi (good sea spirit) 32
Nilsen, A.G. 129
Niyamgiri Mountains 122, 123, 126–7, 128, 130–1, 133–4

Niyamgiri Movement 122, 123–4, 125–6, 127, 129–30, 132–33, 134
Niyamgiri Suraksha Samiti (NSS) 123, 126, 127, 128, 129, 130–1, 132–3
Noongar 25, 26, 27, 28, 34
Norgaard, K.M. 198
normative approach 181
Norström, A.V. 181–2
NSS *see* Niyamgiri Suraksha Samiti (NSS)
Nyikina 27, 28

O

Ober, R. 26
Obrist, B. 191
oceanic literacies 71–72, 74, 83
Odisha *see* Niyamgiri Mountains
Okafor-Yarwood, I. 72
On Country Learning 171, 173
Organization to Save Niyamgiri *see* Niyamgiri Suraksha Samiti (NSS)
'Our house is on fire' (Thunberg) 3
'Our Teachers' protocol 35

P

Pacific Climate Warriors 3, 214, 227
Padel, F. 131
Panchayat (Extention to Schedule Areas) Act (PESA) 114n5
pandemic *see* COVID-19 pandemic
Panthera tigris tigris see Royal Bengal Tiger *(Panthera tigris tigris)*
Paraguassu, Eliete 140
Parsons, M. 11
Particularly Vulnerable Tribe Groups (PVTG) 128
peace, creation of 29–31
pedagogy, decolonizing 163–4, 170–1, 171–3
  see also education
PESA 114n5
Pizarro Choy, A. 12
planetarity 12–13
planetary justice
  about 9–10, 12–13, 180, 239
  solidarity with Country and 172–3
planetary stewardship 55, 91
plurilogues 8, 13–14
Pluriversal Politics (Escobar) 181
Pocock, David 2
politics
  deference politics 143
  prefigurative politics 129, 130
  of solidarity 6–7
polyphonic 6–8
popular intellectuals 125–6, 129, 130, 131, 132–3, 134
possessive logics *see* white possessive logics
postcolonialism 97–8, 101, 110, 111, 114
  see also colonialism; decolonization
potholes 152, 153–4

potion cauldron 233, 234
praxis
  about 143, 147
  coalition-building, of 225
  decolonial 71–2, 83, 90
  ecological 16, 232–3, 236, 238–9
  intersectionality and 100
  oppression and 146–7
  planetary justice and 12
  soil, of 234
prefigurative politics 129, 130
privilege 214
  see also white privilege
profane knowledge 196, 198, 203, 207, 209, 235, 238
prognostic framing 124
  see also collective action frames
projective agency 208
prophecies 26–7
protective silencing 204, 208
Puig de la Bellacasa, M. 237

Q

quality *see* water quality

R

Racial Discrimination Act (RDA) 102–3
racism 219, 223
  see also environmental racism
rainwater harvesting 54, 57–8, 64–5, 66
Ray, S.J. 199
RDA *see* Racial Discrimination Act (RDA)
Redvers, N. 35
regeneration
  collective action 28–9
  decolonization as 31, 35–6
regenerative farming 48–9
Reinvigorate (Hunter) 79, 79–80
religion 62, 146, 168, 185, 223
residential schools 131–2
resilience
  about 180–1
  see also social resilience
resistance 3, 5, 72, 83, 90, 97, 98, 99, 124
responsibility 14, 26, 36, 66, 71, 73, 74, 82, 83, 88
response-ability 26, 31, 36, 36n4, 88, 91
revanchist movement 202–3
rewilding 35
rights
  climate justice and 213–14
  forest rights 98–9, 106–10, 110–11, 112, 114, 115n8
  land rights 98–9, 101–3, 110–11, 113, 213
  river, of 46–7
  sea rights 74
  see also human rights
Rio Tinto 36n3
Rip Curl Pro 73

rivers, rights of 46–7
Robertson, F. 34
Robeyns, I 12–13
Rojeck, C. 103
Roshanravan, Shireen 225
Roy, Arundhati 10
Royal Bengal Tiger *(Panthera tigris tigris)* 63, 179, 184
  see also tigers
Ruddick, Sue 11
Rudolph, Sophie 225
runoff 58, 62
Rutten, R. 125
RWH *see* rainwater harvesting

S

Saad, Layla 225
sacred designs *(miny'tji)* 74, 80
sacred energies 14, 26, 71–2, 73, 76, 84
Sakdapolrak, P. 182
Saltwater Country 71–4, 75, 75–7, 79, 80–4, 82
  see also Country
Saltwater Healing (Hunter) 80, 80
sand art 73, 74–7, 75, 77, 79, 79–80, 80, 81, 82, 83–4
sand talk 73, 83
Sand Talk (Yunkaporta) 34
Sapna South Asian Climate Solidarity 216, 220, 227
Save Narmada Movement *see* Narmada Bacchao Andolan (NBA)
SBR *see* Indian Sundarban Biosphere Reserve (SBR)
scalar tension 12
Scheduled Tribes 106, 114n4
Scheduled Tribes and Other Traditional Forest Dwellers Act *see* Forest Rights Act (FRA)
Schmelzer, M. 203
schooling *see* education
School Strike 4 Climate (SS4C) 34, 213, 215, 218, 220, 221, 223, 224–5, 226
Schrei, J.M. 92
Scotland 196, 201, 202, 205, 207
  see also Learning for Sustainability (LfS)
SDGs *see* Sustainable Development Goals (SDGs)
Sea Country *see* Saltwater Country
sea rights 74
Seed Indigenous Youth Climate Network *see* Seed Mob (youth climate network)
Seed Mob (youth climate network) 3, 34, 214, 227
self-determination 15, 97, 98–100, 101–3, 106–8, 111–12, 113–14
self-governance 107, 114n5
serpent *see* snakes; Waking up the Snake metaphor
settler colonialism 72, 97–8, 101, 103, 110, 111, 114, 144, 163–4, 219

Sharma, Anjali 2, 225, 228n7
shimmer of life 31, 34
Shiva, Vandana 143
silencing *see* protective silencing
Simpson, Leanne 14
Simpson, M. 12
Sixth Assessment Synthesis Report (IPCC) 4
SJSM 185, 188–9
SMART strategies 189, 191
Snow, D. 124
social justice 12–13
social resilience
  about 179, 182
  building of 178, 189–90
  capacities, and 191
soil 233, 234, 235, 236–8, 238–9
solidarity
  ancient wisdom and 25, 26
  anti-coal resistance 105
  bottom-up 16, 67
  capitalism, against 145–8
  class solidarity 7, 140, 142–3, 147–9
  comradeship, and 168–9
  Country, with 169, 171–4
  intersectionality of 112–13, 192, 200, 215
  need, as a 88–9
  networked 180
  politics of 6–7
  situated solidarity 15
  tactical alliances 100, 113
  uneven 220, 226
  youth, and 199–200, 208–9
songlines 28–9, 73, 76–7, 79
songspirals *see* songlines
SOPEC *see* Southern Ocean Protection Embassy Collective (SOPEC)
Southern Ocean Protection Embassy Collective (SOPEC) 73, 78–9
sovereignty
  decolonization and 103
  food 140
  Indigenous justice and 3, 98, 213–14
  pedagogical 163–4
spaces 222, 224, 226–7
  activist 6, 215, 218
  ocean 72–74
  regulated 172
specific-measurable-attainable-relevant-timebound (SMART) strategies *see* SMART strategies
spirits 27, 32, 33, 74, 82, 84, 165–6, 234
Spivak, Gayatri Chakravorty 12–13
Sriprakash, Arathi 219, 225
SS4C *see* School Strike 4 Climate (SS4C)
Standing Rock Sioux 105
Sterlite Industries 127
stewardship, planetary 55, 91
strandings 76–7
Strathern, M. 91

strength, weaknesses, opportunities, threats (SWOT) analysis *see* SWOT analysis
students *see* youth
Sultana, Farhana 11
Sundarban
  region (India) *see* Indian Sundarban Biosphere Reserve (SBR); Indian Sundarban Region (ISR); Sundarban Tiger Reserve (STR)
  Sundarban Jana Sramajibi Mancha (SJSM) 185, 188–9
  Sundarban Tiger Reserve (STR) 183, 184
supremacy *see* white supremacy
sustainability 172
Sustainable Development Goals (SDGs)
  clean water and sanitation (SDG 6) 55, 57
  research and 188
sweet water 58, 59, **61**, 64–5
  see also freshwater
SWOT analysis 188, **189**, 191
Szerszynski, Bronislaw 13

T

tactical alliances 100, 113
Tagore, Rabindranath 159
Táíwò, O. 143
Tatpati, M. 129
Teach the Future (TTF) 201, 203, 205
technological solutions 16, 34, 65
Theriault, N. 35
Thomas, Peter 148
Thunberg, Greta 3, 145
Tide Country *(bhatir desh)* 14, 72
  see also Saltwater Country
tigers 63, 179, 184, 186, 187, 191
tokenism 143, 223
transformative capacities 182, 189
trust 11, 91, 188
TTF *see* Teach the Future (TTF)
Tuck, Eve 5, 7, 72, 167, 168
Turner, B.S. 103
Turtle Island 105, 167
Two-Way Learning 171

U

uneven solidarity 220, 226
UNFCCC 101
UNICEF 57
United Nations
  capitalism, and 141
  Climate Change Conferences (COPs) 2, 12, 80–1
  IPCC 4
  see also Sustainable Development Goals (SDGs)
United Nations Declaration on the Rights of Indigenous Peoples (UNDRIP) 98, 102, 103, 104, 106
United Nations General Assembly 44

United Nations International Children's
　Emergency Fund (UNICEF) 57
United Nations's Framework Convention for
　Climate Change (UNFCCC) 101
United States 105, 166–7, 216
Universal Declaration on the
　RightsofRivers 47

V

vaccine nationalism 2
value *see* intrinsic value
Vedanta Resource Ltd 122, 123–4, 127–8,
　131, 134
Verlie, B. 204, 206, 208
Vikalp Sangam (Alternatives
　Confluence) 130
Villafaña, Dani 215, 220–1, 224–5,
　226–7, 228n5
village councils *see* Gram Sabhas
　(village councils)
Vincent, E. 100, 113
violence
　environmental 14, 16, 76
　Indigenous people, faced by 12, 101,
　　109–10, 111, 134
　intersectional 3–4
　slow, of colonialism 14, 27, 29–31, 144–5,
　　152, 236–7, 239
Viva Energy 73

W

Wadawurrung 73, 74, 76, 77, 84
Waking up the Snake metaphor 27–8, 31–2,
　33, 34, 35, 36
Walker, Catherine 220
Walton, Gloria 224
Wangan and Jagalingou (W&J) Traditional
　Owners 97, 98, 99–100, 103–6, 110,
　111–12, 113, 114, 114n3
Wardan (ocean spirit) 32
Wardandi 25, 26–8, 32, 33
Washington Consensus 141
water *see* fresh water
water, commodification of 41, 58–9, 62–3
water injustice 43, 49, 63, 66
Water, Sanitation and Hygiene (WASH)
　programme 57
water colonialism 43
water injustice 43, 63, 66
water markets 42–3, 49
water quality 43–4, 45–6, 58–60, 62,
　63, 65–6

Webb, Bill 25, 32
West Bengal Drinking Water Sector
　Development Programmes
　(WBDWSDP) 57
Western Sydney 220–1
Western Sydney Landscape Analysis (Australian
　Youth Climate Coalition) 220
Whale Dreaming 76, 77, 77, 81
whales 71, 73, 74, 76–7, 79, 80
whale songlines 73, 76–7, 79
Whanganui Iwi 47
whiteness
　white audacity 214, 219, 222–4
　white consciousness 72, 83, 89
　white innocence 72, 83, 88
　whiteness 200, 214, 215, 216, 218–19,
　　220, 224–5, 228n6
　white possessive logics 222–3
　white privilege 199–200, 201, 214, 219,
　　220–1, 223–5, 228n6
　white supremacy 32, 72, 218–19, 222,
　　223–4, 226, 239
Whyte, Kyle Powys 3, 8, 11, 83, 84
Williams, G. 111, 114
Williams, Lewis 25–6
Wiradjuri 171
wisdom *see* ancient wisdom
W&J *see* Wangan and Jagalingou (W&J)
　Traditional Owners
women
　agriculture and 59
　water access/quality 55, 61–2, 63, 67
working class 15, 146, 147, 148–9
Wray, B. 199
Wretched of the Earth Collective 3, 216

Y

Yang, K. Wayne 5, 7, 72, 167, 168
Yirrkala 80, 83
Yolngu 74
youth
　activism by 34–5, 73, 146–7, 196, 206–7,
　　208–9, 213–14
　BIPOC students 222, 223, 224–5
　on education 201–4
　emotions 197–200
　see also School Strike 4 Climate (SS4C)
Yunkaporta, Tyson 34, 83, 88
Yusoff, Kathryn 9

Z

Zapatistas 132

www.ingramcontent.com/pod-product-compliance
Lightning Source LLC
Chambersburg PA
CBHW051533020426
42333CB00016B/1905